A Colour Atlas
of *Bacillus* Species

Published by Wolfe Medical Publications Ltd 1983
Printed by W.S. Cowell Ltd, 8 Buttermarket, Ipswich, England
ISBN 0 7234 0777 0

General Editor, Wolfe Medical Books
G Barry Carruthers MD(Lond)

This book is one of the titles in the series of Wolfe Medical Atlases, a series which brings together probably the world's largest systematic published collection of diagnostic colour photographs.

For a full list of Atlases in the series, plus forthcoming titles and details of our surgical, dental and veterinary Atlases, please write to Wolfe Medical Publications Ltd, Wolfe House, 3 Conway Street, London W1P 6HE.

A Colour Atlas of *Bacillus* Species

Jennifer M. PARRY
FIMLS
Senior Medical Laboratory Scientific Officer
Bacteriology Department
School of Pathology
Middlesex Hospital Medical School, London

P.C.B. TURNBULL
BSc, MS, PhD
formerly Principal Grade Microbiologist
Food Hygiene Laboratory
Central Public Health Laboratory
Colindale, London

J.R. GIBSON
Head of the Department of Medical Illustration
Central Public Health Laboratory
Colindale, London

Wolfe Medical Publications Ltd

Contents

Acknowledgements

The production of this atlas could never have been possible without the generous and enthusiastic assistance of a very large number of people. The authors are greatly indebted to these persons.

We are particularly grateful to Dr L.R. Hill, Curator of the National Collection of Type Cultures, Central Public Health Laboratory, Colindale, for his invaluable help in assembling the cultures used and final reading of the manuscript. We also thank the Culture Collections listed on page 253 for the supply, free of charge, of a number of the cultures used in the atlas.

We are most grateful to Dr H.D. Burges, Insect Pathology Group, Glasshouse Crops Research Institute, Littlehampton, West Sussex, and Dr R.P. Williams, Department of Microbiology and Immunology, Baylor College of Medicine, Texas Medical Center, Houston, for particular aid with data for the manuscript and for useful criticisms of relevant sections of it during its preparation.

We have appropriately credited in the text those persons or concerns who kindly supplied us with cultures, photographs, information or the use of their facilities; we acknowledge our gratitude to all these persons and concerns.

In addition, we thank the following for various types of assistance: Dr H. de Barjac, Institut Pasteur, Paris; Mr B.J. Capel and Mr J.A. Carman, PHLS-CAMR, Porton Down, Salisbury; Dr R.Y. Cartwright and Mr C.F. Daines, Public Health Laboratory, Guildford; Dr B.N. Dancer, Department of Biochemistry, Oxford University; Dr L. Davies, National Institute for Research in Dairying, Shinfield, Reading; Mr D.E. Gray, Librarian, Central Veterinary Laboratory, Weybridge and Mr J.F. Spilsbury, Curator, Wellcome Bacterial Collection, Beckenham.

We are greatly indebted to the South African Institute for Medical Research, Johannesburg, for many aspects in facilitating preparation of the manuscript; in particular, the extensive coverage of *B. anthracis* would not have been conceivable without the assistance and facilities of that Institute and its staff.

Abbott Laboratories, South Africa (Pty) Ltd are acknowledged with gratitude for bearing costs of postage.

Finally, the project would not have been possible without the support and encouragement of the Public Health Laboratory Service and of the Director, Dr R.J. Gilbert, and the staff of the Food Hygiene Laboratory, in particular, Mr J.M. Kramer. We appreciate this support and encouragement.

Photography

The majority of the pictures in this volume were taken in the Department of Medical Illustration, Central Public Health Laboratory, Colindale, London, on a Nikon F2 camera using a Nikkor 55 mm Macro lens with either extension tubes or a bellows unit to obtain the desired magnification.

The camera was attached to a copying stand and lighting of the Petri dishes for ×1 pictures and of biochemical reactions was by 500 watt photo-flood lamps. Lighting for ×6 magnification of colonies was by fibre-optic illumination from a halogen light source (Volpi, Switzerland).

The film used was almost always Kodak-Ektachrome 50 ASA Tungsten with Process E6.

Figures 21, **23** and **24** are from 6 × 7cm Agfa 50L transparency originals taken by Mr M.G. Anderson, Photographic Unit, South African Institute for Medical Research, Johannesburg.

Spore photographs were taken on a Zeiss photomicroscope at the Institute of Orthopaedics, Royal National Orthopaedic Hospital, Stanmore, Middlesex, by kind courtesy of Mr D. Sayer.

Preface

Although *Bacillus* species are frequently encountered in clinical laboratories, generally speaking, little effort is made to identify them. This is primarily because of (1) the rather too ready acceptance that they are contaminants or saprophytes and of no clinical consequence, and (2) the dauntingly enormous range of colonial morphologies that arises between not just different species, but even between strains of the same species.

However, a relatively recent interest in several members of the *Bacillus* group as potential pathogens has resulted in more laboratories wishing to identify them.

The Food Hygiene Laboratory, Colindale, by reason of its interest in *B. cereus* as a food poisoning agent, inadvertently became a centre to which increasing numbers of all types of aerobic spore-bearers were sent for identification, not just from foods and food poisoning incidents, but also from non-food related infections and other environmental sources.

Clinical laboratories frequently need to be able to decide rapidly on the preliminary identification of an isolate and not infrequently we received requests for guidelines on identification in the case of *Bacillus* species.

The first encounter the clinical laboratory worker makes with a *Bacillus* isolate is usually the colonies on a blood agar plate. It was our experience that colonial morphology on blood agar was of considerable help in the tentative identification of the isolate and a valuable pointer to the subsequent tests that needed to be done to confirm the identification.

The idea of a photographic colour atlas orientated around the type and range of colonial morphologies exhibited by *Bacillus* species arose from the feeling that it would supply a welcome aid in clinical laboratories to the rapid preliminary identification of members of the genus *Bacillus*.

In the form in which it has ultimately evolved, the volume comprises colour plates showing the colonial morphologies of a wide variety of common *Bacillus* species together with a selection of lesser known species. Guidelines to follow-up identifications are based on simple cell, spore and biochemical characteristics such as can easily be tested or looked for in a routine clinical laboratory.

Deeper aspects of taxonomy, such as G + C estimations of the DNA, have only been referred to superficially for interest purposes. The principal aim is that the atlas should be a practical guide to the busy bench bacteriologist with only simple tests to hand. No claim is made to it being a fundamental taxonomic authority; rather it is a bench demonstration of cultures that have been identified and classified according to the most widely accepted manual on *Bacillus* species (Gordon *et al*, 1973) and recognized culture collection strains have been used to the fullest possible extent in its compilation. The project was made possible by access to strains from such culture collections through the kind assistance of the National Collection of Type Cultures, also situated at Colindale. The authors are greatly indebted to Dr L.R. Hill, Curator of the National Collection, for this assistance.

In addition to species descriptions, an attempt has been made throughout to outline the many features among the various members of the genus of interest to clinical, agricultural, pharmaceutical, food, dairy and other industrial microbiologists. Although in all of these fields except clinical bacteriology blood agar is not a commonly used medium, it is hoped that the volume will nevertheless be of interest and value to microbiologists in all these areas.

Introduction

Such attempts as had been made before about 1950 to identify aerobic sporing bacilli to species level had only left the taxonomy of this large group of bacteria in a state of great confusion.

The preface to the USDA Agriculture Handbook No. 427, 'The Genus *Bacillus*' (Gordon *et al*, 1973) outlines the events and the work which culminated in the publications of the USDA Miscellaneous Publication No. 559, 'Aerobic Mesophilic Sporeforming Bacteria' (Smith *et al*, 1946) and the USDA Monograph No. 16, 'Aerobic Sporeforming Bacteria' (Smith *et al*, 1952). These, for most persons interested in *Bacillus* taxonomy, marked the stages at which some order was brought into the taxonomy of the *Bacillus* group.

There would be little dispute either that Handbook No. 427 itself, which replaced the earlier monographs, became and remains the principal authority on the taxonomy of this genus although the contributions of other workers and publications such as that of Wolf and Barker (1968) should not go unappreciated.

The genus *Bacillus* embraces a large number of bacteria with a great diversity of properties. Proposals have been made from time to time to simplify the genus by dividing it into several genera, but each of the proposed arrangements has had its problems (Gibson and Gordon, 1974) and the group has remained intact.

Smith *et al* (1946, 1952) found they were able to divide these aerobic spore-bearers into three groups on the basis of the shape of the spore and the swelling or absence of swelling of the sporangium by the spore. They recognized, however, that these groups did not represent strictly ordered, distinct compartments but rather that, within each group, there existed a spectrum of properties. In the same manner as a colour has a particular place within the spectrum of light, so a *Bacillus* species appeared to these workers to have a particular place within the spectrum of its group.

For example, the spectrum of Morphological Group 1, defined by absence of sporangial swelling and possession of ellipsoidal spores, begins with *B. megaterium* which is followed by *B. cereus, B. licheniformis, B. subtilis, B. pumilus, B. firmus* and finally *B. coagulans*. The cells of these species were found, on the whole, to be progressively smaller in diameter and, with the exception of *B. coagulans*, their spores did not appreciably distend the sporangia. The spores of *B. coagulans* may or may not swell the sporangium and this species was seen as an intermediate or link between this first group and Morphological Group 2 which comprises species whose sporangia are swollen by oval spores. The Group 2 species were, in turn, arranged across a spectrum of decreasing activity on carbohydrates.

Morphological Group 3 was defined to encompass species forming round spores; however, although the authors clearly felt that other species existed within this group, only in the case of *B. sphaericus* did they feel the species were adequately represented to allow specific descriptions.

The order in which the *Bacillus* species were listed by Gordon *et al* (1973) was thus based on what these authors judged to be the most appropriate position for each species within these spectra. That the subdivision into three groups on the basis of spore shape and sporangial swelling has its limitations was pointed out by Wolf and Barker (1968) who proposed a modified scheme. However, apart from minor exceptions, the order and arrangement given by Gordon *et al* (1973) has been adhered to in this atlas; even the minor differences, such as the positions of 'B. pulvifaciens' and *B. pasteurii* have been based on qualifications within the text of Gordon *et al*.

There remained also a number of 'unclassified' (Smith *et al*, 1952) or 'unassigned strains/poorly represented taxa' (Gordon *et al*, 1973); the authors felt their experience with these was too limited to enable them to define their taxonomically useful properties or to prepare complete species descriptions. In the eighth edition of *Bergey's Manual*, Gibson and Gordon (1974) classified these as *'incertae sedis'* on the basis that each species was at present represented by just one or two or only a very few strains and that more evidence was needed to establish their status as distinct species.

Finally, in the latest taxonomic summary of *Bacillus* species drawn up by the Subcommittee on the Taxonomy of the genus *Bacillus* (International Committee on Systematic Bacteriology) and published in the Approved Lists of Bacterial Names (1980), some 18 of the unassigned *'incertae sedis'* species were not included. These, then, from a strict taxonomist's point-of-view, join a 'sizeable total' (Gordon, 1981) of unidentified strains of *Bacillus* in the various culture collections that exist.

It is perhaps not easy for the non-taxonomically minded bacteriologist to understand why it should be so difficult to give names to these strains. The idea of a 'spectrum' is again helpful in grasping the problem involved – this time as applied to just a single species. As can be readily seen simply in the colonial morphologies illustrated in this atlas, even the well-known species of *Bacillus*, such as *B. subtilis*, *B. licheniformis* and *B. cereus*, are not composed of uniform and homogeneous groups of strains. Rather, 'the component strains are relatively heterogeneous and form a graduated series of differences' (Gordon *et al*, 1973). The unidentified *Bacillus* strains fall beyond the accepted extremes of the spectra of the defined species. Suitable tests revealing patterns of identity between the many odd unidentified species or between them and the accepted species have just not been found. Furthermore, the ability to adapt to a wide range of different environments appears to be a characteristic of some *Bacillus* species and it thus becomes essentially impossible to define a set of tests that are appropriate for all *Bacillus* species from all environments.

The problems for the *Bacillus* taxonomist and some examples are reviewed by Gordon (1981). Essentially, the taxonomic experts prefer to have a relatively small number of well-defined species and leave the remainder as 'unidentified' until better taxonomic tools become available than to have a long string of species represented by just one or two known strains. Dr Gordon tells us that 'More strains of the poorly represented taxa are needed so that it may be established how many of the unclassified strains typify distinct species' (Gordon, 1981). It is certainly one hope that the atlas will stimulate the collection of further strains so that fuller classification of the *Bacillus* species can be achieved. The clinical laboratory must rate as one of the best potential sources of such strains. And for this reason, representatives of the unassigned groups have been included in the atlas using the nomenclature of Gordon *et al* (1973).

In the meantime, excellent work is in progress (Berkeley and Goodfellow, 1981) which is using novel taxonomic methodology to determine the relationships between the unidentified strains and the well-established species and to extend the methods that are available for identification at a practical level. Hopefully a future edition of the atlas may be feasible which will reflect and complement the results of this work.

As the work which ultimately led to this book progressed, the question 'which species should be included in a book which belongs to a clinical reference series?' did arise. It was, however, hard to answer. While, in moments of clinical enthusiasm, a number of *Bacillus* species other than *B. anthracis* have been associated to one extent or another with infections, it remains the general rule in the average routine laboratory to dismiss aerobic spore-bearers, when encountered, as contaminants. There is evidence (Tuazon *et al*, 1979; Gilbert *et al*, 1981; Gordon, 1981) that members of this group of bacteria should be taken more seriously as potential pathogens of man and animals.

While *B. cereus* stands out particularly in this respect, and to a lesser extent, *B. licheniformis* and *B. subtilis*, it has been our feeling that the clinical significance of other species within the genus may be underestimated and would become better recognized if 'ASBs' (aerobic spore-bearers) could be more readily identified when encountered.

As a group, *Bacillus* species are ubiquitous in nature. Due to the resistance of their spores to desiccation, heat, cold, disinfectants and other destructive forces, they persist through a range of adverse conditions which would quickly kill other bacteria. For this reason, they are indeed frequent contaminants of 'sterilized' materials, surfaces, foods, culture media and so on. But for the same reason, they are often in a position second to no other bacterial group to enter wounds or to be injected, inhaled or ingested.

While it is not our intention to overplay the potential pathogenic roles of *Bacillus* species, it is our contention that, until the matter is studied in a more thorough and consistent manner, the relevance of the frequent encounters by the clinical laboratory worker with ASBs will never be properly assessed. It is hoped that this atlas will enable such laboratory workers to approach ASBs in that more consistent manner.

Although methods, media and supplementary sections on clinical and commercial aspects of certain of the *Bacillus* species are included, the body of the atlas is comprised of figures showing colonies of representative strains of the species in each of the morphological groups. Recognized culture collection strains have been used where possible; no differences in colonial morphology were observed between culture collection strains and fresh isolates of those species we encountered.

The term 'representative' needs to be stressed; within the species from which many strains were available for examination, such as *B. subtilis, B. cereus* and *B. licheniformis*, wide variation in colonial morphology is found. On the other hand, despite this variation, it was felt that colonies were very often tentatively recognizable as belonging to a particular species. In the case of species for which very few strains are yet available for examination, it seems reasonable to expect the same conditions of variety and identity to apply.

Such an assumption, while commensurate with one of the fundamental premises around which the atlas has been prepared – namely that colonial morphology is useful in the identification of *Bacillus* species – is open to criticism as unproven. This, however, only increases our hope that this volume will not only serve to assist in the practical identification of members of the better established species but will also lead to the finding and collection of more representative strains of the rarer species which can then be more completely defined.

With the exception of *B. popilliae*, the illustrations of colonial morphology show it on blood agar – the basic plating medium of the clinical laboratory. It is recognized that other types of laboratories with an interest in *Bacillus* species may not use blood agar and that colonial morphologies can differ greatly on different agar media. Nevertheless, it is hoped that by providing an overview of the genus *Bacillus* which has not hitherto been readily available to the general microbiological public, the atlas will be of interest and value to bacteriologists in the food, dairy, pharmaceutical, agricultural and other fields concerned with this group of bacteria.

Methods and characterization tests

I. General culture methodology
II. Incubation and storage
III. Microscopical examination
  1. Gram staining
  2. Spore morphology
  3. Parasporal crystals
    (a) Malachite green method
    (b) Buffalo black (naphthol blue-black) method
    (c) General information on parasporal crystals of *B. thuringiensis*
  4. Motility
IV. Biochemical tests
  1. Lecithovitellin/Lecithinase production (LV reaction)
  2. Citrate utilization
  3. Anaerobic growth
  4. Voges-Proskauer (V-P) reaction
  5. Nitrate reduction
  6. Indole production
  7. Growth in 7% sodium chloride
  8. Starch hydrolysis
  9. Casein hydrolysis
  10. Gelatin hydrolysis (liquefaction)
  11. Urease production
  12. Ammonium salt sugars
V. A demonstration of isolation and identification as exemplified in the isolation of *B. cereus* from rice
VI. Additional tests for more specific purposes
  1. Haemolysis
  2. Propionate utilization
  3. Decomposition of tyrosine
  4. Growth in presence of 0.001% lysozyme
  5. Deamination of phenylalanine
VII. Typing systems
  1. General
  2. Serotyping of *B. cereus*

I. General culture methodology

In probably the majority of cases, the clinical microbiologist will be faced with identifying the principal *Bacillus* species in Morphological Groups 1, 2 and 3. **Figure 1** is a flow chart giving suggested primary steps in the identification of these species. **Table 1** lists secondary identification characteristics and the final biochemical identification tests are given in **Tables 2–5**. These tests are based on those described by Gordon *et al* (1973) and included by Cowan (1974).

Figure 1 Primary steps in the identification of principal *Bacillus* species of Morphological Groups 1, 2 and 3.

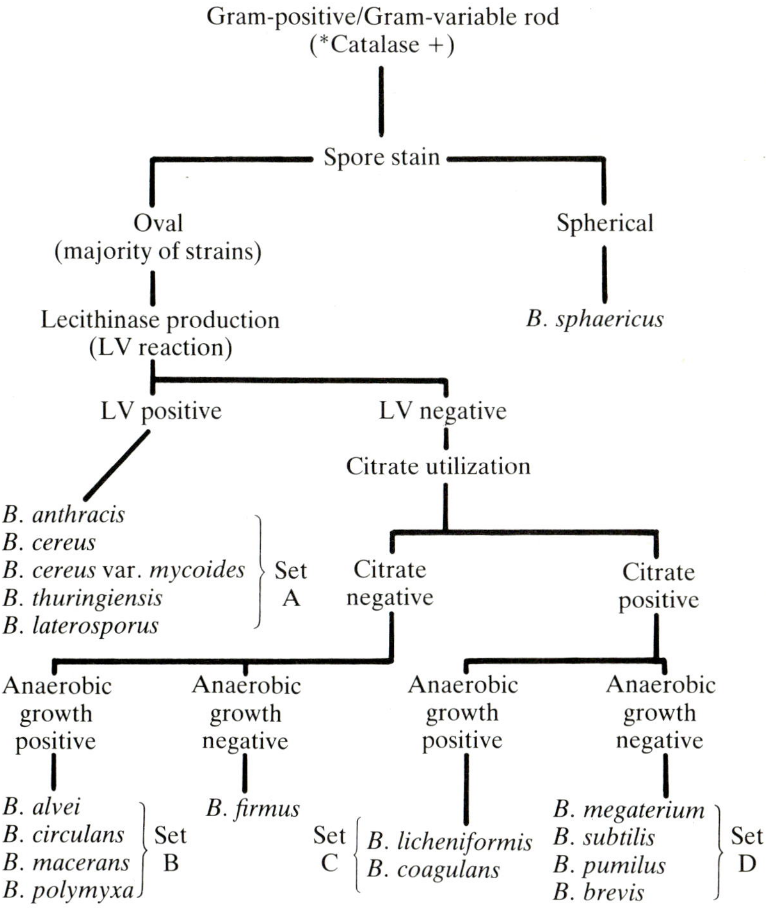

*B. *larvae* and *B. popilliae* are catalase negative and 'B. pulvifaciens' is weakly positive only.

Table 1 Secondary characteristics in the identification of the principal *Bacillus* species in Morphological Groups 1, 2 and 3.

Set A LV positive	
B. anthracis	– non-motile, non-haemolytic, Morph. Group 1.
B. cereus	– marked haemolysis, motile, Morph. Group 1.
B. thuringiensis	– identical to *B. cereus* contains parasporal bodies in young culture.
B. cereus var. *mycoides*	– variant of *B. cereus*, colonies rhizoid and spreading. Usually non-motile.
B. laterosporus	– typical canoe-shaped cells containing Morph. Group 2 spores.
Set B LV negative, citrate utilization negative, grow anaerobically	
B. alvei	– typical motile microcolonies on plate surface. V-P positive, fails to reduce nitrate.
B. circulans	– V-P negative, nitrate reduction variable. Anaerogenic in carbohydrate media.
B. macerans	– V-P negative, reduces nitrate, produces gas in carbohydrates.
B. polymyxa	– V-P and nitrate reduction positive, large mucoid colonies often produced.
Set C LV negative, citrate utilization and anaerobic growth positive	
B. licheniformis *B. coagulans*	– differentiated by gelatin hydrolysis – *B. licheniformis* hydrolyses gelatin while *B. coagulans* does not.
Set D LV negative, citrate utilization positive, anaerobic growth negative	
B. megaterium	– V-P negative, nitrate reduction variable. Morph. Group 1.
B. subtilis	– V-P and nitrate reduction positive. Morph. Group 1.
B. pumilus	– V-P positive, fails to reduce nitrate. Morph. Group 1.
B. brevis	– this is the only Morph. Group 2 species in this set. V-P negative, variable nitrate reduction.
Other strains	
B. firmus	– V-P negative, nitrate reduction positive.
B. sphaericus	– round spores, non-fermentative, hydrolyses gelatin.

Table 2 Biochemical characteristics of Morphological Group 1 species.

	B. megaterium	*B. cereus*	*B. cereus* var. *mycoides* (*B. mycoides*)	*B. anthracis*	*B. thuringiensis*	*B. licheniformis*	*B. subtilis*	*B. pumilus*	*B. firmus*	*B. coagulans*
LV (egg yolk) reaction	−	+	+	+	+	−	−	−	−	−
Citrate utilization	+	+	+	v	+	+	+	+	−	v
Anaerobic growth	−	+	+	+	+	+	−	−	−	+
V-P reaction	−	+	+	+	+	+	+	+	−	v
Nitrate reduction	v	+	+	+	+	+	+	−	+	v
Indole production	−	−	−	−	−	−	−	−	−	−
Growth in 7% NaCl	+	+	+	+	+	+	+	+	+	−
Starch hydrolysis	+	+	+	+	+	+	+	−	+	+
Casein hydrolysis	+	+	+	+	+	+	+	+	+	v
Gelatin hydrolysis	+	+	+	+	+	+	+	+	+	−
Urease activity	v	v	v	−	v	v	v	−	−	−
Ammonium salt sugars acid from:										
glucose	+	+	+	+	+	+	+	+	v	+
mannitol	v	−	−	−	−	+	+	+	+	v
xylose	v	−	−	−	−	+	+	+	v	v
arabinose	v	−	−	−	−	+	+	+	v	v
Haemolysis (blood agar)		+	+	−	+	(see text)				
Motility		+	−	−	+	(see text)				
Propionate utilization						+	−			
Parasporal bodies		−	−	−	+					
Tyrosine hydrolysis	+/−	+	+/−	−	+	−	−	−	−/+	−
Growth in 0.001% lysozyme	−	+	+	+	+	−	−/+	+/−	−	−

v = variable; +/− = more often +; −/+ = more often −.

Table 3 Biochemical characteristics of species in Morphological Groups 2 and 3.

	Morph. Group 2									Morph. Group 3	
	B. polymyxa	*B. macerans*	*B. circulans*	*B. stearothermophilus*	*B. alvei*	*B. laterosporus*	*B. brevis*	'B. pulvifaciens'	*B. popilliae*	*B. sphaericus*	*B. pasteurii**
LV (egg yolk) reaction	−	−	−	−	−	+	−	−	−	−	−
Citrate utilization	−	−	−	−	−	−	v	−	−	v	−
Anaerobic growth	+	+	v	−	+	+	−	+	+	−	+
V-P reaction	+	−	−	−	+	−	−	−	−	−	−
Nitrate reduction	+	+	v	+	−	+	v	+	−	−	+
Indole production	−	−	−	−	+	v	−	−	−	−	−
Growth in 7% NaCl	−	−	v	−	−	−	−	−	+	−	−
Starch hydrolysis	+	+	+	+	+	−	−	−	−	−	−
Casein hydrolysis	+	−	v	−	+	+	+	+	−	v	−
Gelatin hydrolysis	+	+	+	v	+	+	+	+	−	+	+
Urease activity	−	−	−	−	−	−	−	−	−	v	+
Ammonium salt sugars											
acid from:											
glucose	+	+	+	+	+	+	v	+	+	−	−
mannitol	+	+	+	v	−	+	v	+	−	−	−
xylose	+	+	+	v	−	−	−	−	−	−	−
arabinose	+	+	+	v	−	−	−	−	−	−	−
Growth at 65°C				+							

v = variable.
*1% urea must be added to all media.

Table 4 Biochemical characteristics of the unassigned strains of subgroups A, B and C.

	Subgroup A				Subgroup B			Subgroup C			
	'B. apiarus'	'B. filicolonicus'	'B. thiaminolyticus'	*B. alcalophilus*	'B. cirroflagellosus'	'B. chitinosporus'	*B. lentus*	*B. badius*	'B. aneurinolyticus'	'B. macroides'	'B. freundenreichii'
LV (egg yolk) reaction	−	−	−	−	−	−	−	−	−	−	−
Citrate utilization	+	−	−	−	−	−	−	−	−	−	−
Anaerobic growth	+	+	+	+	−	−	−	−	−	−	−
V-P reaction	−	−	−	−	−	−	−	−	−	−	−
Nitrate reduction	+	+	+	−	+	−	−	−	+	v	−
Indole production	−	−	+	−	−	−	−	−	−	−	−
Growth in 7% NaCl	+	+	−	−	−	−	v	+	−	v	+
Starch hydrolysis	+	+	+	+	+	+	+	−	−	−	−
Casein hydrolysis	+	+	+	+	−	+	−	+	−	v	−
Gelatin hydrolysis	+	+	+	+	−	+	−	+	−	+	−
Urease activity	+	−	+	−	−	−	+	−	−	−	+
Ammonium salt sugars acid from:											
glucose	+	+	+	−	−	+	+	−	−	v	−
mannitol	−	+	−	−	−	−	v	−	−	v	−
xylose	−	−	−	−	−	−	+	−	−	−	−
arabinose	−	v	v	−	−	−	v	−	−	v	−
Phenylalanine deamination					+	−	v	−	−	+	+

v = variable.

Table 5 Biochemical characteristics of the unassigned strains of subgroups D, E_1 and E_2.

	Subgroup D		Subgroup E_1				Subgroup E_2	
	B. pantothenticus	'B. epiphytus'	'B. aminovorans'	*B. globisporus*	*B. insolitus*	'B. psychrophilus'	'B. psychrosaccharolyticus'	*B. macquariensis*
LV (egg yolk) reaction	−	−	−	−	−	−	−	−
Citrate utilization	−	v	−	−	−	−	−	−
Anaerobic growth	+	−	−	−	−	−	+	+
V-P reaction	−	−	−	−	−	−	−	−
Nitrate reduction	+	+	v	v	−	+	+	−
Indole production	−	−	−	−	−	−	−	−
Growth in 7% NaCl	+	+	v	−	−	v	v	−
Starch hydrolysis	v	+	+	v	−	+	+	+
Casein hydrolysis	+	+	−	v	−	v	+	−
Gelatin hydrolysis	+	−	−	+	−	+	+	−
Urease activity	−	−	+	+	−	+	−	+
Ammonium salt sugars acid from:								
glucose	+	+	+	+	−	+	+	+
mannitol	−	+	−	−	−	v	+	+
xylose	−	+	−	−	−	v	+	+
arabinose	−	−	−	−	−	−	+	−
Phenylalanine deamination	+	+	−	+	v	−	−	−

v = variable.

In keeping with the intention that this should be a reference book for clinical laboratories, all strains used for preparing this atlas were initially plated on 5% horse blood agar (Oxoid Columbia base in our laboratory), incubated at 37 ± 1°C and examined at 24 hours and, if necessary, at 48 and 72 hours also.

Blood agar was not appropriate for the insect pathogen, *B. popilliae*, and would not have been for the fairly closely related insect pathogen, *B. larvae*, which has not been included in this atlas.

II. Incubation and storage

The majority of strains were mesophilic; unless otherwise stated, incubation for growth and biochemical tests was carried out at 37 ± 1°C and inocula for biochemical and physical tests were taken from colonies on the blood agar plates.

In the case of the psychrophilic strains (unassigned strains in subgroups E_1 and E_2), the blood agar plates were incubated at 22°C for 2–3 days.

The thermophilic species, *B. stearothermophilus*, was incubated at 60°C for 24 hours. *B. coagulans*, occasionally encountered in food samples for example, is also relatively thermophilic and has an optimum growth temperature of 45–50°C (Kramer *et al*, 1982). The plates of *B. coagulans* shown in **Figures 73–75**, however, were incubated at 37°C as for the generally mesophilic strains and this temperature is perfectly satisfactory for the culture of this species.

B. pasteurii required a 1% urea supplement in all media including nutrient agar storage slopes. We also added this supplement to the Christensen's urea medium when testing *B. pasteurii* for its ability to produce urease.

All strains except *B. popilliae* were stored on nutrient agar slopes at room temperature.

Aerobic spore-bearers isolated in the clinical laboratory will almost always be those that grow readily (i.e. are apparent within 24–48 hours) in or on a general nutrient medium at 37°C. Nevertheless, an attempt has been made to cater for the widest possible range of purposes for which this atlas may be consulted and to allow for the variations in species characteristics which exist within the genus as a whole. Thus, while on account of their fastidious growth requirements *B. larvae* and *B. popilliae* would not normally be seen in a clinical laboratory, it was felt that the inclusion of one of these would be of interest while adding to the comprehensiveness of the atlas.

Tryptone–glucose–yeast extract broth ('J' medium), which satisfies the thiamine requirements of *B. larvae* and *B. popilliae*, was used for growth (25–30°C in a shaking water bath) and maintenance of the strain of *B. popilliae* (**Figure 109**). For maintenance, subculture to fresh 'J' medium was done every month. For demonstration of the colonies, a 2-week-old culture in diphasic 'J' medium was plated on to 'J' agar and the colonies were photographed after 48 hours' incubation at 22°C.

Similarly, the clinical laboratory microbiologist is, for the most part, concerned with growth and test results that can be read after 24 or 48 hours of incubation at 37°C. For many of the *Bacillus* species and the tests that are described below, these conditions would prove sufficient. Nevertheless, in the descriptions and instructions that are given for these tests, it has been necessary to qualify '37°C' with such statements as 'or other temperature if known to be more appropriate'. Incubation periods have frequently had to be given in terms of the outside time at which a negative reaction can be accepted as truly negative.

Some flexibility of thought is, therefore, necessary in following the guidelines given below.

III. Microscopical examination

Gram and spore stains should be done on all isolates.

2

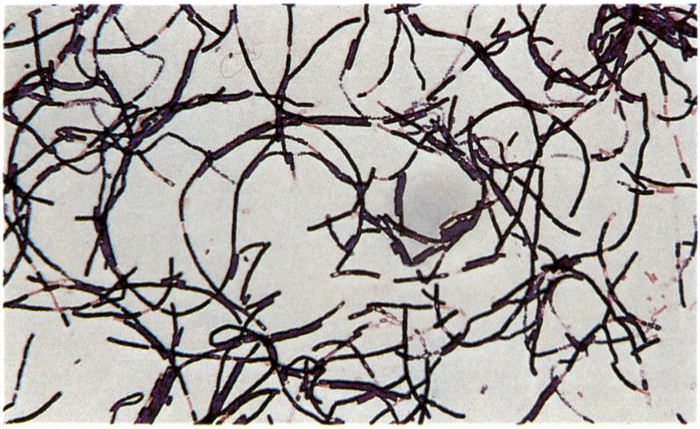

Figure 2 Gram stain. Staining has been carried out on a 48 hour (37°C) culture of *B. anthracis* A.19432/80 on blood agar. Note the cells or groups of cells where Gram positivity has been lost. × 400.

III. 1. Gram staining

Although generally thought of as Gram-positive rods, Gram variability is quite commonly encountered among the aerobic sporeformers and, on occasion, strains are found which can only be described as Gram-negative.

Gram positivity may be lost with age (**Figure 2**) and where there is Gram variability, the cells are Gram-positive in the early stages of growth (Gibson and Gordon, 1974). Gram staining should, therefore, be carried out as soon as is feasible – i.e. usually within 24 hours of plating out and incubating on blood agar.

Procedure (Modification of the method of Lillie, 1928)

(i) Heat fix an air-dried thin smear of the organisms on a microscope slide.
(ii) Flood the slide with ammonium oxalate–crystal violet for 30 seconds.
(iii) Wash with running tap-water (see precautions given in section VIII. 1 if the culture is suspected of being *B. anthracis*).
(iv) Flood the slide with Lugol's iodine for 30 seconds.
(v) Tip off the iodine, decolourize with a few drops of acetone and wash immediately under running tap-water.
(vi) Counterstain with 0.5% aqueous safranin for 30 seconds.
(vii) Wash well under the tap and blot dry.

Gram-positive organisms are blue-black or dark purple; Gram-negative bacteria are red.

III. 2. Spore morphology

For most routine purposes, blood agar plate cultures left on the bench for 2–4 days are adequate for sporulation and observation of spore morphology. Occasionally strains are encountered that do not spore readily on blood agar and sporulation agar or soil extract agar is required. Sporulation medium is also necessary periodically to confirm that certain isolates are not *Bacillus* species.

Procedure (Schaeffer and Fulton, 1933)

(i) Heat fix an air-dried thin smear on a microscope slide.

(ii) Flood the slide with 5% aqueous malachite green and steam for 1 minute (e.g. with spirit burner).

(iii) Wash away all the stain with running tap-water (if *B. anthracis* is suspected, wash into phenolic disinfectant – see section VIII. 1).

(iv) Counterstain with 0.5% aqueous safranin for 15–45 seconds.

The bacterial cells (sporangia) stain red while the spores are green. The shape and position of the spore and any swelling of the sporangium are looked for (**Figure 3**).

These features may alternatively be looked for using phase contrast microscopy and a number of examples are shown later in this book (**Figures 39, 103**).

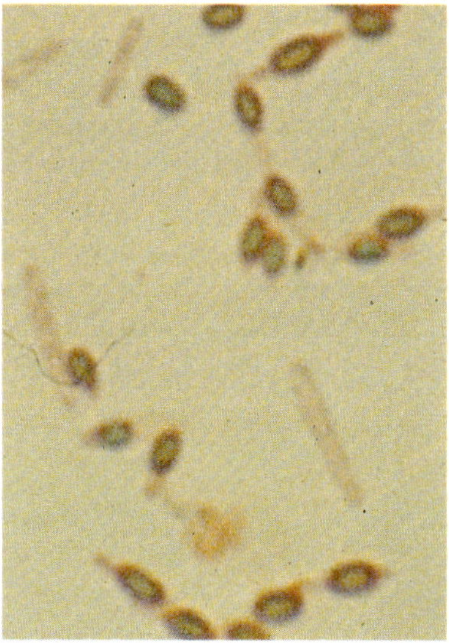

3

Figure 3 Spore staining and morphology. Malachite green stain of spores of *B. brevis*. Note how the sporangia (faint red rods) are definitely swollen by the spores which are central to subterminal. × 1600.

III. 3. Parasporal crystals

The most likely reason that users of this atlas will search for parasporal bodies will be to differentiate *B. thuringiensis* from *B. cereus*. *B. popilliae* also produces a refractile parasporal body and it has recently been found that a few strains of *B. sphaericus* also produce parasporal bodies (Dr H.D. Burges, personal communication).

The crystals can be observed intracellularly in the sporangium by phase contrast microscopy of sporulation agar cultures or using a staining procedure. The following staining procedures are suitable.

III. 3. a. Malachite green method

This is a modified method of the spore stain procedure given above (III. 2).

(i) Heat fix an air-dried smear of a sporulated culture with the spores and crystals released from the sporangia (the spores and crystals are released naturally by autolysis of the sporangia during the second and third days of growth on sporulation agar).
(ii) Flood with 5% aqueous malachite green and steam gently until the edges of the pool of stain begin to dry.
(iii) Add more stain and repeat (ii) twice.
(iv) Wash away all green with tap-water, blot off excess water and cover with dilute (0.1%) carbol fuchsin for 30 seconds.
(v) Wash under running tap-water and blot dry.

The spores stain green (as in **Figure 3**) and the crystals deep pink. The pink-staining crystals are often cuboidal or bipyramidal appearing as square or diamond-shaped in these smears. Square-shaped crystals can be seen in **Figure 4A** and **C** and diamond-shaped ones are clear in **Figure 4B**.

4

A B

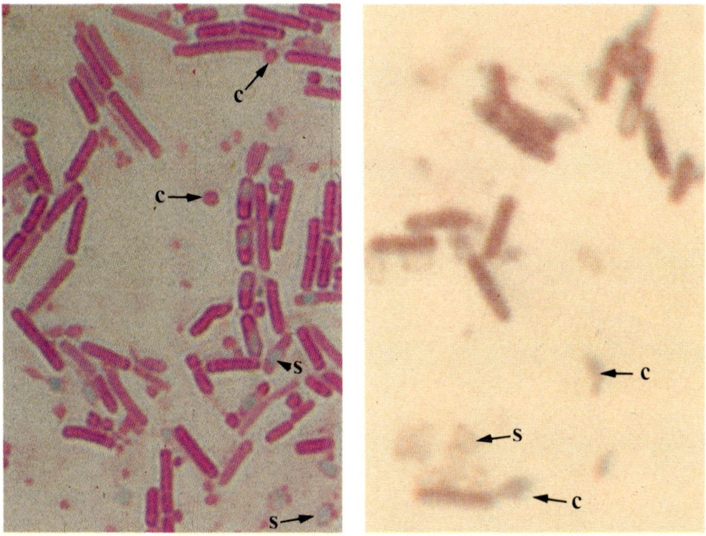

Figure 4 Parasporal crystal bodies *(B. thuringiensis).* **A** Malachite green stain. × 1800. ***B** Buffalo black (naphthol blue-black) stain. × 1250. s = spore; c = crystal.

III. 3. b. Buffalo black (naphthol blue-black) method (**Figure 4B**)

(i) Heat fix an air-dried smear of a sporulated culture with spores and crystals released from the sporangia as per III. 3. a.
(ii) Cover with a thin film of buffalo black stain while the slide is still warm from heat fixing. Leave for 70 seconds.
(iii) Wash the stain off with cold tap-water, blot off excess water and cover with full strength (1%) carbol fuchsin for 20 seconds.
(iv) Wash again under running tap-water, blot and allow to dry completely. Examine under oil immersion.

The spores stain pink around the periphery; the crystals stain black (**Figure 4B**).

*The stained preparation was kindly supplied by Dr H.D. Burges.

4

C

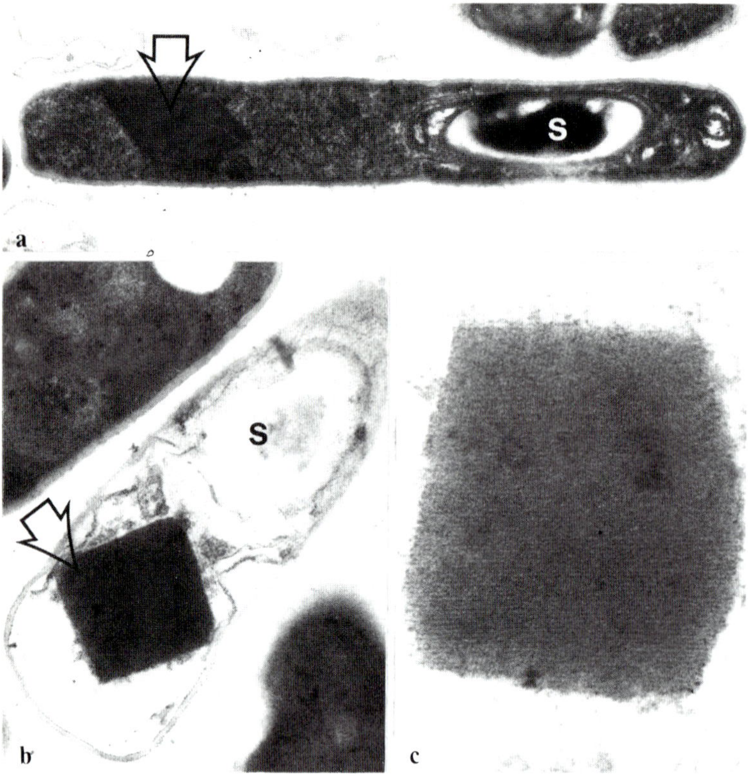

Figure 4 C Parasporal crystals of *B. thuringiensis*.

Electron micrographs of thin-sectioned sporulated cells of *B. thuringiensis* subsp. *tolworthi* from a culture on blood agar (24 hours at 37°C followed by 24 hours at room temperature). (**a**) Relatively young cell, ×16 700; sporangium still 'healthy' but developing spore (S) and crystal body (arrow) are readily visible. (**b**) Later stage, ×22 500; sporangium approaching lysis; crystal body fully formed. (**c**) crystal body, ×77 600; showing lattice structure. (Electron microscopy by Dr I. Popiel, Department of Tropical Pathology, South African Institute for Medical Research, Johannesburg.)

III. 3. c. General information on parasporal crystals of B. thuringiensis

The parasporal crystals of *B. thuringiensis* are formed at the same time as, but apparently independently of, the spores (Bechtel and Bulla, 1976).

These crystal bodies are of considerable economic and agricultural interest and importance due to their insecticidal properties. They are composed of a single glycoprotein of apparent molecular weight of 134000 which is a 'pro-toxin'. Under exposure to the alkaline conditions and enzymes of the larval insect's midgut, this is converted to a molecule which is actively toxic to susceptible insects.

B. thuringiensis isolates appear to be divisible into those whose toxins are most toxic to lepidopterous larvae (moths in particular) and those whose crystal proteins are most toxic to dipterous larvae (mosquitoes and black flies) (Tyrell *et al,* 1981).

The crystal lattice structure can be seen in **Figure 4C (c)**.

B. thuringiensis viable spores mixed with parasporal crystals comprise the active ingredients of commercial insecticides used extensively in several countries of the world (Burges, 1981).

Of medical interest is the claim that the crystal protein is capable of bringing about regression in tumours and of enhancing the overall immune response in mammals (Prasad and Shethna, 1974, 1975).

III. 4. Motility

Although various tests for motility are available, in our experience, the two that are most reliable are growth through semi-solid agar (in a Craigie tube) and the hanging drop method.

For the hanging drop, young (4–7 hours) nutrient broth cultures incubated at 30°C should be examined in conventional hanging drop preparations using a high power objective or phase contrast if available.

A pair of Craigie tubes can be seen at top right of **Figure 20**. The inner tube is inoculated from a colony on a blood agar plate using a straight wire. After 24–48 hours of incubation at 30°C, organisms that have 'swum' to the outside surface of the agar can be picked off for plating or inoculation of further broths as required. The method is useful for collection of the most motile members of a population of bacteria as, for instance, required in preparation of flagellar antigen for serotyping (see section VII and **Figure 20**).

It is often heard that the hanging drop method is the most reliable test for motility; however, Brown *et al* (1958) found all of 115 strains of *B. cereus* were motile when tested in semi-solid medium while only 97 of these had been judged motile by hanging drop examination.

Figure 5 illustrates the peritrichous flagella around a cell of *B. cereus* which are responsible for its motility.

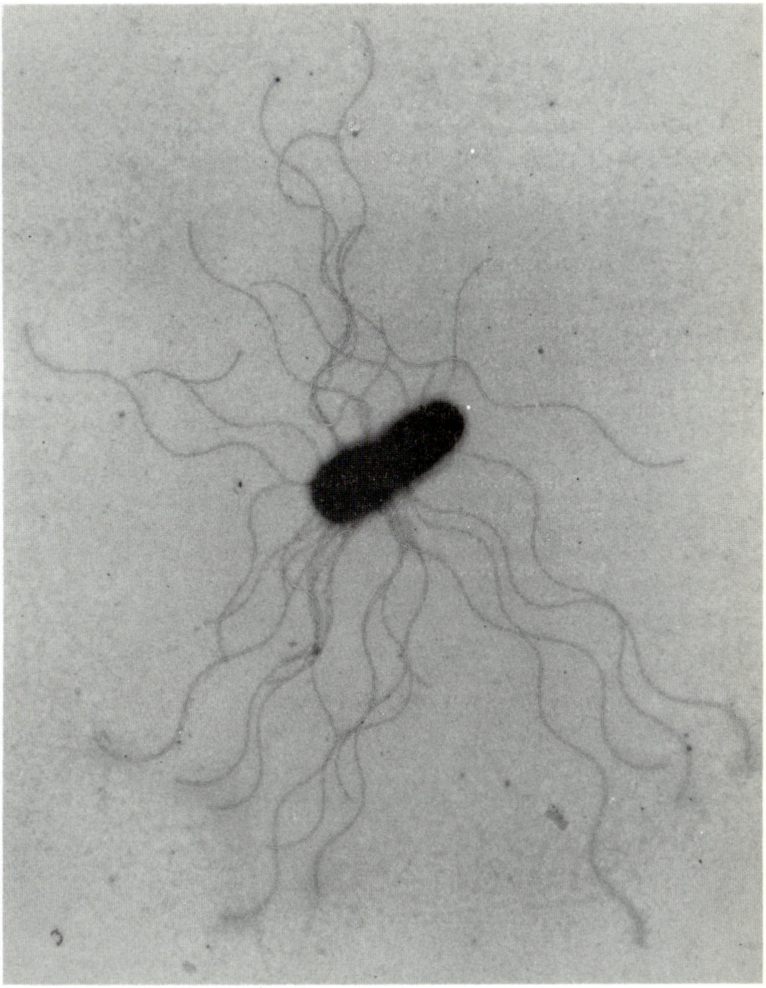

5

Figure 5 Peritrichous flagella around *B. cereus* F4433/73 ×6200. (Electron microscopy by Dr I. Popiel, Department of Tropical Pathology, South African Institute for Medical Research, Johannesburg.)

IV. Biochemical tests

IV. 1. Lecithovitellin/Lecithinase production (LV reaction – **Figure 6**)
The scheme given in **Figure 1** is based on the experience that the most common organism we are required to confirm or identify is *B. cereus*. Detection of lecithovitellin production was thus a primary test. The test utilizes the hydrolysis of the phosphatid lecithin of egg yolk (Colmer, 1947) by the as yet imprecisely defined action of phospholipase-C.

Kendall's BC medium (**Figure 6**) was designed in this laboratory (Miss M. Kendall) to facilitate recognition of positive LV reactions in preliminary identification of *B. cereus*. Other egg yolk based media which are readily to hand in the average clinical laboratory, such as that of McClung and Toabe (1947) or Willis and Hobbs (1958, 1959) as used for the Nagler reaction, are generally acceptable also, however.

Procedure

Strains under test are stabbed, streaked (**Figure 6A**) or smeared (**Figure 6B**) on to the egg yolk medium which is incubated at 37°C for 24 hours and examined for zones of opalescence around the culture growth.

6

A 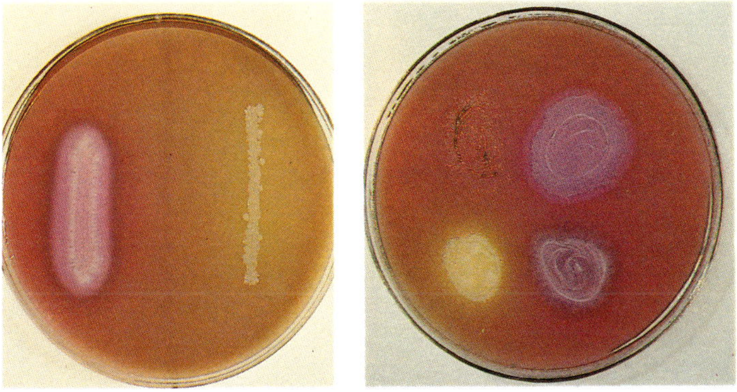 B

Figure 6 **Lecithovitellin/lecithinase production (LV reaction).**
24 hours: 37°C.

A Cultures streaked on to Kendall's BC medium. Left: Mannitol −/LV +. Right: Mannitol +/LV −.

B Cultures smeared on to Kendall's BC medium. Top right: Mannitol −/strong LV +. Lower right: Mannitol −/weak LV +. Lower left: Mannitol +/LV −. Top left: Mannitol −/LV −.

Kendall's medium serves the dual role of making the zone of opalescence more readily visible and demonstrating fermentation (or its absence) of mannitol. A change from purple to yellow indicates positive fermentation of mannitol. The mannitol fermenter in **Figure 6A** has turned the medium more colourless than yellow.

Knight and Proom (1950) noted that a number of *Bacillus* species other than those listed in **Figure 1** under LV positive give a reaction with egg yolk, but one which is restricted to the medium immediately below the culture growth. They referred to this as 'egg-yolk reaction of the "restricted type"'. In looking for the LV positive strains of **Figure 1**, therefore, the plates should be read without scraping the growth off.

Species recorded as giving egg yolk reactions of the restricted type are *B. alvei, B. brevis* (weak), *B. lentus* (weak), *B. macerans* (weak), *B. polymyxa* (variable and weak when positive) and *B. pumilus* (weak).

A commercially produced medium similar to Kendall's BC medium is available (Oxoid CM617).

7

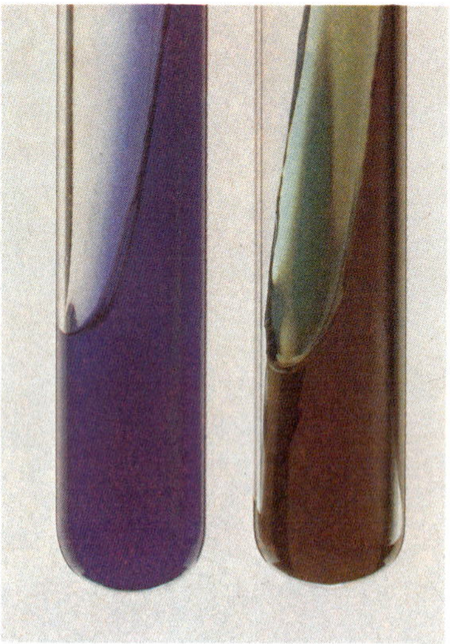

Figure 7 Citrate utilization. Left = +. Right = −.

IV. 2. Citrate utilization **(Figure 7)**

(i) Make a light suspension of the organism picked from the surfaces of colonies on a blood agar plate.
(ii) Inoculate a Simmons' citrate slope with a stab and a single streak using a straight wire.
(iii) Incubate at 37°C (or other temperature if known to be more appropriate) for up to 2 weeks.
(iv) Utilization of citrate is indicated by development of a deep purple (alkaline indicator colour) from the original green in the medium.

(Caution: carry-over of the medium on or in which the organism has been grown may lead to invalid results (Cowan, 1974). A straight wire rather than a loop is therefore recommended for inoculation of the citrate slope, and the inoculation should be made from a light suspension of the culture in water, saline or buffer as indicated in (i).)

8

Figure 8 Anaerobic growth in BBL anaerobic agar. Left = +. Right = −.

IV. 3. Anaerobic growth **(Figure 8)**

Gordon *et al* (1973) were satisfied with anaerobic agar (BBL) for determination of the ability of *Bacillus* strains to grow anaerobically. For several years this was used in the Food Hygiene Laboratory although it has now been displaced by the conventional anaerobic jar technique (Kramer *et al*, 1982).

In the anaerobic agar method, cultures are inoculated to the bottom of tubes of the agar which are incubated at 37°C (or other temperature if known to be more appropriate) for 1–7 days. The tubes are examined for obvious growth along the line of the stab (**Figure 8**). The method is rapid and satisfactory for most purposes.

For strict anaerobic testing, blood agar plates, pre-reduced by holding anaerobically for 24 hours before use, should be spread and incubated at the optimum growth temperature for at least 48 hours in an anaerobic jar containing an Eh indicator (BBL) or control culture of *Pseudomonas aeruginosa*.

9

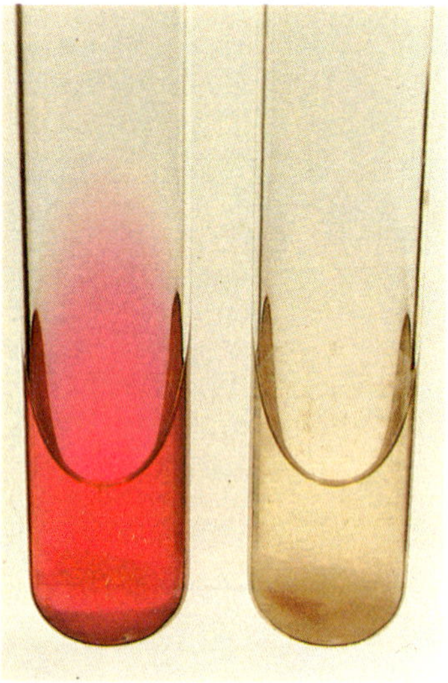

Figure 9 Voges–Proskauer (V-P) reaction. Broth cultures in two tubes tipped at an angle. Left = +. Right = −.

IV. 4. Voges–Proskauer (V-P) reaction **(Figure 9)**

Two tubes of glucose–phosphate broth should be inoculated from the blood agar culture and incubated at 37°C (or other temperature if known to be more appropriate). The first tube may be tested for the presence of acetoin at 2–3 days; if negative, the second tube should be tested at 7 days.

In this laboratory, V-P tubes contain 5 ml of glucose–phosphate broth; acetoin production is tested for by addition of 0.6 ml of 5% α-naphthol solution followed by 0.2 ml of 40% potassium hydroxide solution. The tube is shaken and left at a slant (for aeration).

Development of a red colour within 15 minutes is taken as a positive reaction.

IV. 5. Nitrate reduction **(Figure 10)**
In its simplest form, the test for nitrate reduction consists of inoculating nitrate broth and testing after 2 or 3 days of incubation at 37°C for nitrite production.

Stricter testing involves incubating nitrate broth cultures for up to 2 weeks. This may be done by inoculating more than one tube and testing them at appropriate periodic intervals. Alternatively (Gordon *et al*, 1973), strips of potassium iodide starch paper (Gallard–Schlesinger Chemical Mfg. Corp., Carle Place, New York) may be moistened with a few drops of 1N hydrochloric acid. At each testing period, one of these is touched with a loopful of the nitrate broth culture. Development of a purple colour indicates the presence of nitrite. At the end of 2 weeks, or at the point at which the KI-starch paper gives a positive reaction, the nitrite test should be carried out on the culture.

The nitrite test consists of adding 0.5 ml of each of nitrate reagents A and B to the nitrate broth culture. The development of a red colour within 1–2 minutes indicates the presence of nitrite.

Occasionally organisms reduce nitrate beyond nitrite. Addition of a small amount of zinc dust to a tube appearing negative after addition of reagents A and B will reduce any residual nitrate to nitrite. Thus a true negative will now turn red, while a broth in which the nitrate has been reduced beyond nitrite will not develop a red colour. All negatives should be checked with zinc dust.

Gas production under normal conditions is not a characteristic of *Bacillus* species; under anaerobic conditions, however, certain Morphological Group 1 species reduce nitrate to nitrogen gas. This may be relevant in *B. cereus* gas gangrene (Fitzpatrick *et al*, 1979).

10

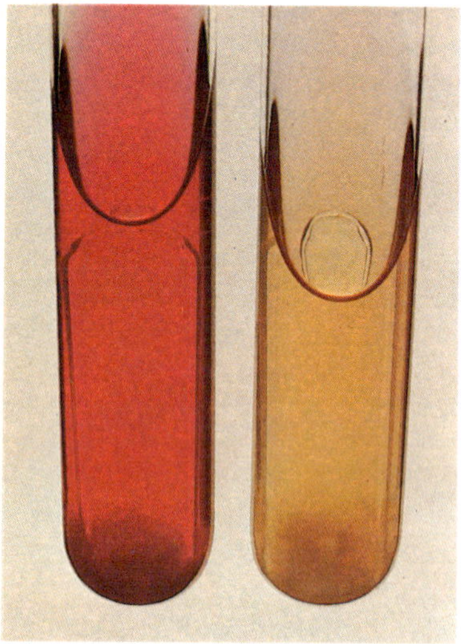

Figure 10 Nitrate reduction. Broth cultures in two tubes tipped at an angle. Left = +. Right = −.

11

Figure 11 Indole production.
Left = +. Right = −.

IV. 6. Indole production **(Figure 11)**
In its simplest form, 0.5 ml of Kovac's or Ehrlich's reagent is added to a 24 or 48 hour culture of the test strain in peptone water or tryptone broth. The mixture is shaken. A rapid development of a red colour in the alcoholic layer which settles out on top indicates the presence of indole and thus a positive indole test.

As with citrate, V-P and nitrate tests above, it may frequently be wise not to accept a negative reaction in cultures incubated for less than 2 weeks. A simple approach is to inoculate more than one peptone water tube; if at 24 or 48 hours, the reaction is negative, the test can be repeated after longer incubation times.

12

Figure 12 Growth in 7% salt trypticase broth. Broth cultures in two universal bottles tipped at an angle.

Left – turbid; growth has occurred.

Right – clear; no growth has occurred.

IV. 7. Growth in 7% sodium chloride **(Figure 12)**

In this laboratory, 7% (w/v) salt trypticase broth is used; nutrient broth containing 7% salt (w/v) is equally good.

The broth is inoculated with the strain and turbidity (i.e. growth) is looked for after incubation at 37°C (or other growth temperature if known to be more suitable) for up to 14 days.

13

Figure 13 Starch hydrolysis. Left = +. Right = −.

IV. 8. *Starch hydrolysis* (Figure 13)

A loop touched to a colony of the strain on blood agar is streaked lightly across a potato-starch plate which is then incubated at 30°C for 4 days.

The plate is then flooded with a 1:5 dilution of Lugol's iodine solution. Unchanged starch turns blue-black; hydrolysis of starch is indicated by a clear zone around the streak (**Figure 13**).

An alternative to Lugol's iodine is 95% ethanol. In this case, the unchanged starch becomes white and opaque within 15–30 minutes; again a clear zone around the streak indicates starch hydrolysis and the plate appears similar to that seen in **Figure 14**.

Weakly positive strains may not produce a visible zone around themselves and, in the case of apparent negatives, some of the growth should be scraped off to see whether there is a clear zone underneath the culture growth.

14

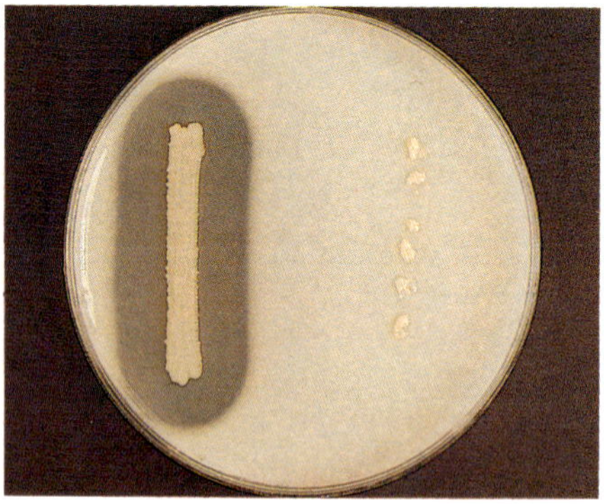

Figure 14 Casein hydrolysis. Left = +. Right = −.

IV. 9. Casein hydrolysis (**Figure 14**)

A casein agar plate is streaked with a loop previously touched to a colony of the strain under test. The plate is incubated at 37°C (or other temperature if known to be more appropriate) for 1–14 days.

Clearing around the streak (or beneath the growth in the case of weakly positive strains) is looked for and indicates decomposition of the casein.

15

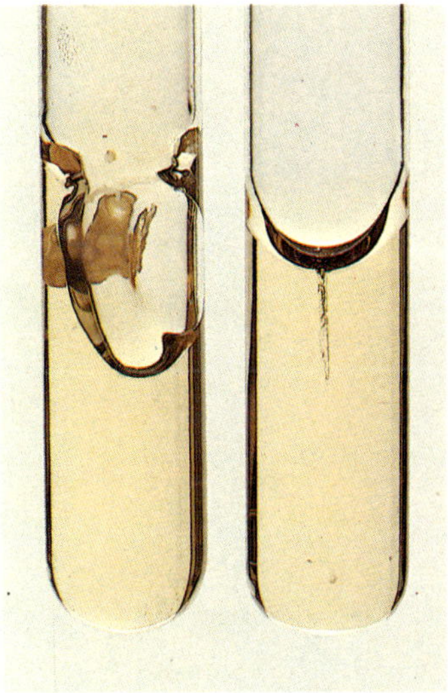

Figure 15 Gelatin hydrolysis (liquefaction). The tubes have been tipped at an angle.

Left: +; the liquefied gelatin has flowed along the tipped tube.

Right: −; the gelatin has remained solid and has not flowed along the tipped tube.

V. 10. Gelatin hydrolysis (liquefaction) **(Figure 15)**

noculate a tube of nutrient gelatin by stabbing with a straight wire that ıas been touched to a colony of the strain under test on blood agar. ncubate at 20–22°C and observe daily for growth and liquefaction.

Alternatively, incubation may be done at a higher temperature Gordon *et al*, 1973, give 28°C), but, before reading, the tubes should e held at 20°C for about 4 hours to allow unchanged gelatin to harden.

The tube should be incubated for 4 weeks before a negative result is ccepted.

16

Figure 16 Urease production. Left = +. Right =

IV. 11. Urease production **(Figure 16)**

Stab a Christensen's urea slope with a heavy inoculum picked up on a straight wire from a colony of the strain under test on blood agar.

Incubate at 37°C (or other temperature if known to be more appropriate).

Examine daily up to 14 days for development of the magenta colour indicating that hydrolysis of urea (by urease) has occurred.

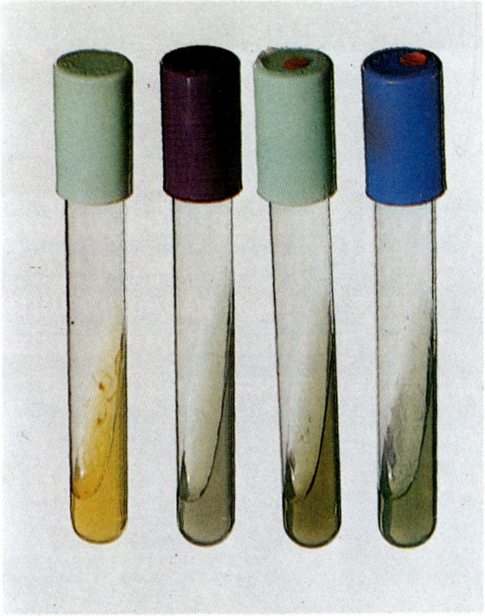

17

Figure 17 Ammonium salt sugars.
The 3 tubes on the right – green = –.
Tube on the left – yellow = +.

IV. 12. Ammonium salt sugars **(Figure 17)**
The use of an ammonium salt plus yeast extract base in carbohydrate tests for *Bacillus* species has evolved from the results of a number of studies. Smith *et al* (1946) emphasized that, because of their strong proteolytic nature, many of these bacilli may not form enough acid from a carbohydrate within an ordinary peptone-containing base medium to overcome the alkalinity produced from organic nitrogen during proteolysis, and thus may not effect a change in the indicator even though the sugar is actually fermented. They suggested that the base should consist instead of a synthetic agar medium containing only inorganic nitrogen (0.1% ammonium phosphate) to which the carbohydrate under test was added to a final concentration of 0.5%.

Knight and Proom (1950), however, found that a number of *Bacillus* species needed other nutritional substances (amino acids and/or aneurin, biotin, nicotinic acid or pantothenic acid). The formula for the carbohydrate base was thus modified (Smith *et al*, 1952) to include 0.02% yeast extract.

It was not universally accepted that the ammonium salt plus yeast extract base was ideal for differential carbohydrate tests. Burdon (1956), for instance, found that a distinctive difference between *B. licheniformis* and *B. subtilis* in fermentation of maltose in a tryptose agar base was lost when the ammonium salt plus yeast extract base was used.

Nevertheless, it has become conventional today to follow Gordon *et al* (1973) in their use of the ammonium salt plus yeast extract base for carbohydrate fermentation tests.

Procedure

A set of glucose, arabinose, mannitol and xylose ammonium salt sugar slopes (and other carbohydrates as required) should be inoculated with a loop previously touched to a colony of the strain under test on blood agar.

The slopes should be incubated at 37°C (or other temperature if known to be more appropriate) and observed for 5–14 days for acid production (green to yellow – **Figure 17** – in the media used in this laboratory).

Gas bubbles throughout the agar should be noted as indicatory of gas production; this is especially important in the final identification of *B. circulans* and *B. macerans*. Norris *et al* (1981) imply that ammonium salt sugars in liquid form with inverted Durham tubes are perhaps more suitable for detecting gas production where this is essential.

V. A demonstration of isolation and identification as exemplified in the isolation of *B. cereus* from rice (Figure 18)

Figure 18 is a diagrammatic flow chart illustrating isolation and identification of a *Bacillus* species *(B. cereus)* using some of the tests of section IV at the simplest level.

The food in the honey jars is rice and *B. cereus* is very commonly found in rice. It is also readily recognized on blood agar plates, is easily picked off for purity and is characteristically LV + and mannitol − on Kendall's BC medium. It is glucose + in ammonium salt sugars and negative in mannitol, arabinose and xylose, and salicin variable.

In this laboratory, motility is established as part of the serotyping procedure (**Figure 20**) through which all *B. cereus* isolates are put. Haemolysis is noted on the initial blood agar isolation and purity plates.

In general this is sufficient for the identification of *B. cereus* once the laboratory worker is familiar with it. It would be ideal if parasporal crystals were looked for routinely since we simply do not know how often what is identified as *B. cereus* is, in fact, *B. thuringiensis*.

For less easily identified *Bacillus* species, the whole set of tests in sections III and IV above should be applied methodically as indicated in **Figure 1** and **Tables 1–5**.

18

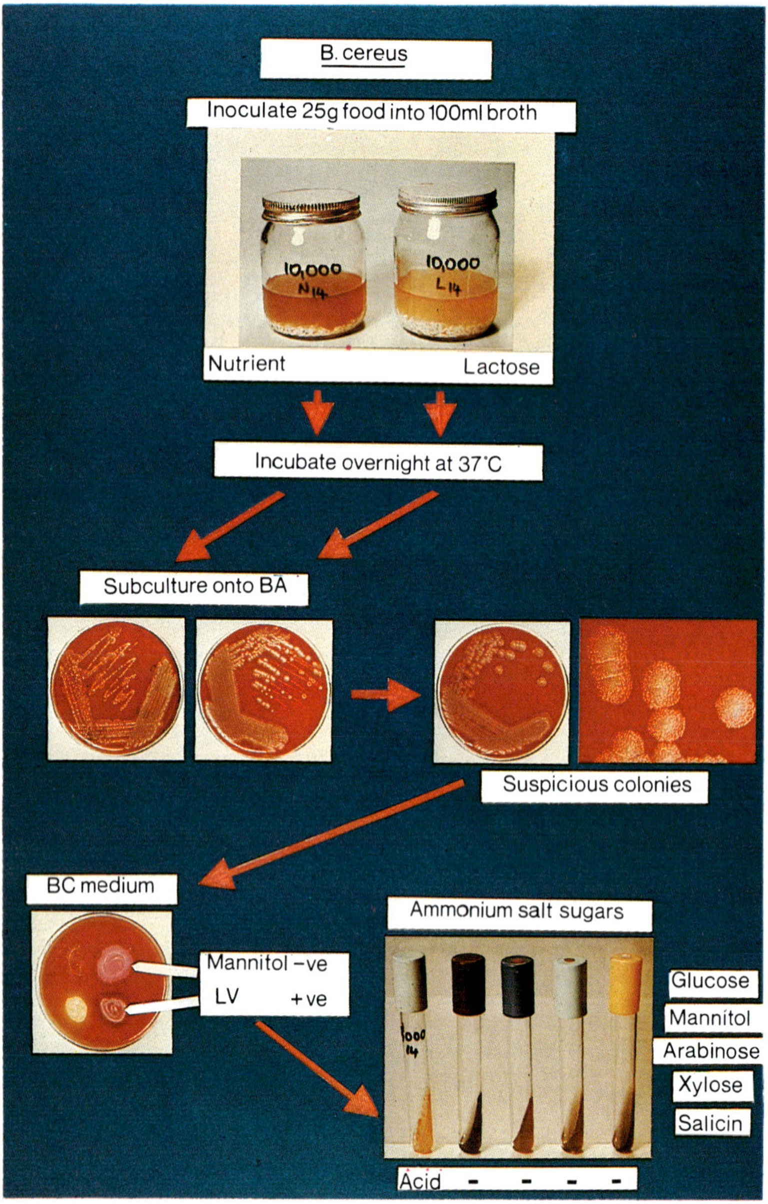
B. cereus
Inoculate 25g food into 100ml broth
10,000 N14
10,000 L14
Nutrient
Lactose
Incubate overnight at 37°C
Subculture onto BA
Suspicious colonies
BC medium
Mannitol −ve
LV +ve
Ammonium salt sugars
Glucose
Mannitol
Arabinose
Xylose
Salicin
Acid - - - -

VI. Additional tests for more specific purposes

The following tests are useful in providing further identification aids for individual or restricted groups of *Bacillus* species.

VI. 1. Haemolysis

Inspection for zones of haemolysis around colonies on sheep or horse blood agar is a preliminary guide in the differentiation of *B. anthracis* from *B. cereus* (see section VIII. 4).

Typical *B. cereus*-like colonies which are not haemolytic on blood agar plates that have been incubated for 24 hours at 37°C and which yield non-motile bacilli should be handled with caution. However, great variation exists in the extent to which *B. cereus* strains exhibit haemolysis and lack of apparent haemolysis should only be regarded as indicating a need for caution.

Figure 18.

19
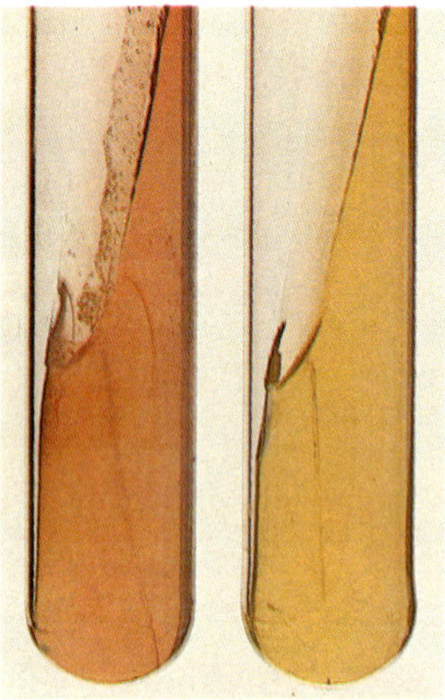

Figure 19 Propionate utilization. Left = +. Right = −.

VI. 2. Propionate utilization **(Figure 19)**

This test is particularly useful in differentiating *B. subtilis* (−) from *B. licheniformis* (+) which are, in many respects, very alike.

The medium consists of the same base as Simmons' citrate agar except that 0.2% sodium propionate replaces the 0.2% sodium citrate in the latter.

A different indicator (in this case, phenol red) is used in our laboratory to distinguish the two tests more readily, but development of an alkaline reaction (indicated by a change from yellow to pink) signifies propionate utilization.

As with the citrate test (section IV. 2), the inoculum should be taken with a straight wire from a light saline suspension of the organism and not directly from a colony on an agar plate. A single stab and streak should be used, with incubation at 37°C (or other temperature if known to be more appropriate) for up to 2 weeks.

VI. 3. Decomposition of tyrosine

Tyrosine agar plates are inoculated with a streak of the organism under test in the same manner as with starch or casein agar plates (section IV. 8, 9, **Figures 13**, **14**).

The plates are incubated at 37°C and observed for up to 2 weeks for the development of zones of clearing (hydrolysis) around or beneath the culture growth.

Gordon *et al* (1973) evidently found this test useful in the differentiation of a number of Morphological Group 1, 2 and 3 strains and several unassigned strains.

In particular, they listed this as a differential characteristic for *B. anthracis* and *B. cereus*.

VI. 4. Growth in the presence of 0.001% lysozyme

A tube of 0.001% lysozyme in nutrient broth is inoculated with the strain being tested and incubated at 37°C (or other temperature if known to be more appropriate). The tubes are examined at regular intervals up to 14 days for turbidity indicating the ability of the strain to grow in this medium.

VI. 5. Deamination of phenylalanine

Tubes of malonate broth are inoculated with a loop touched to a colony of a strain under test. After 24 hours' incubation at 37°C (or optimum temperature of strain under test), the culture is acidified with 0.1–0.2ml of 0.1N HCl. Then 0.2ml of 10% (w/v) ferric chloride is added, and the tube shaken and examined for a colour change. A positive reaction is indicated by an immediate green colour that quickly fades.

VII. Typing systems

VII. 1. General

Typing systems have been developed for a number of *Bacillus* species, primarily as taxonomic tools with which to establish relationships between strains or species or to subspeciate species, but also for epidemiological purposes in demonstrating a relationship (or lack of it) between isolates from different sources following epidemiological or ecological investigations.

Useful references to work on development and applications of bacteriophage, bacteriocin and serological typing systems for various of the *Bacillus* species can be found in a publication by Berkeley and Goodfellow (1981) by consulting these topics in their index.

The best example of the use of serotyping for subspeciating a species is in the classification of the insect pathogen, *B. thuringiensis* into different serotypes as based on flagellar (H) antigen agglutination (de Barjac and Bonnefoi, 1962; de Barjac, 1981). On the basis of serotyping in conjunction with biotyping, there are 19 accepted subspecies of *B. thuringiensis*. *B. sphaericus*, another insect pathogen, has similarly been subspeciated on the basis of a flagellar (H) antigen serotyping method (de Barjac, 1981).

VII. 2. Serotyping of **B. cereus** **(Figure 20)**

H antigen serotyping systems have been developed for *B. cereus* by Lemille *et al* (1969) and also in this laboratory (Taylor and Gilbert, 1975; Gilbert and Parry, 1977; Kramer *et al*, 1982) and have been found to be useful in the epidemiological investigations of food poisoning and other infections by *B. cereus*.

Not all isolates of *B. cereus* can be typed with the present set of 23 antisera in this laboratory's scheme, but further antisera are continually being assessed for the potential value they might offer if added to the existing set.

Figure 20 shows the *B. cereus* H serotyping procedure used in this laboratory.

Pools of sera are used to detect initially whether the strain is typable and it is frequently not necessary to dilute all the way to the end point. Thus, depending on the particular needs in view, smaller volumes of the nutrient broth culture than are shown may be used.

The culture should also be allowed to stand for an hour or so after addition of the formalin to allow this to fix the cells.

Incubation of the agglutination tray or tubes may be 50°C for 2 hours as shown or overnight at 37°C.

(Note: the number 10000 on the tubes and flask is simply a mock specimen number. The '14' and '15' on the Craigie tubes are dates illustrating that to select the most motile organisms and thus optimal levels of flagellar (H) antigen, the culture was passaged through 2 Craigie tubes on 2 consecutive days.)

20

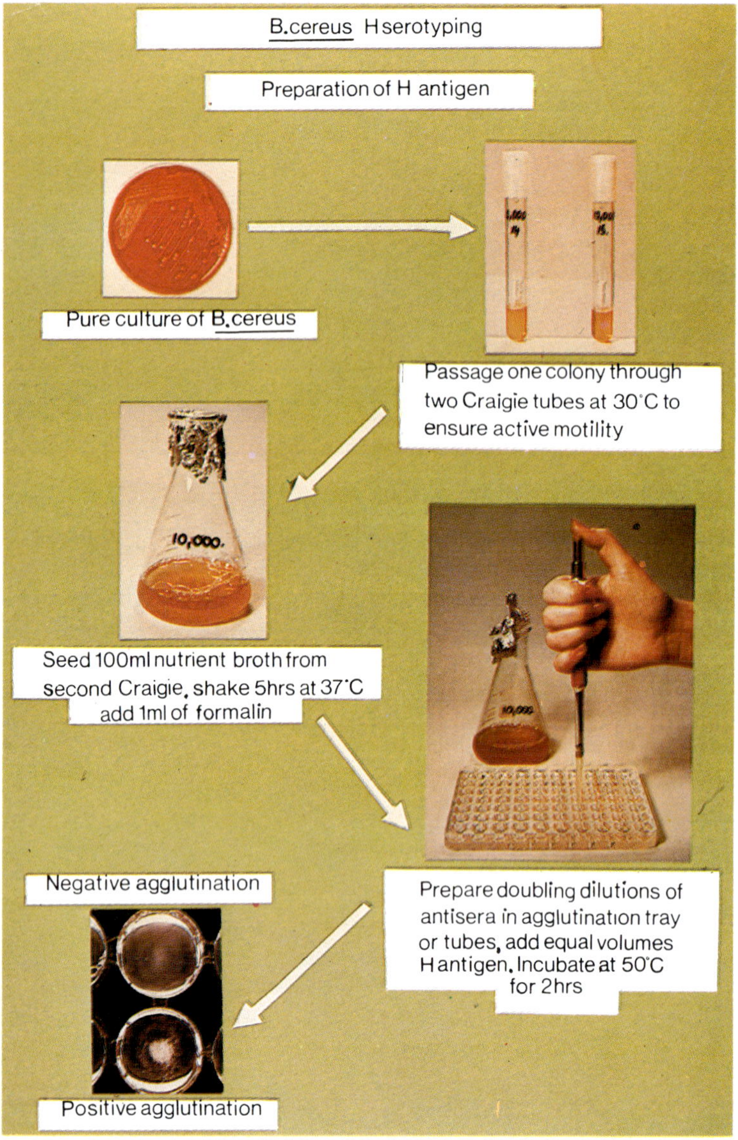
B.cereus H serotyping
Preparation of H antigen
Pure culture of B.cereus
Passage one colony through two Craigie tubes at 30°C to ensure active motility
10,000.
Seed 100ml nutrient broth from second Craigie, shake 5hrs at 37°C add 1ml of formalin
Prepare doubling dilutions of antisera in agglutination tray or tubes, add equal volumes H antigen. Incubate at 50°C for 2hrs
Negative agglutination
Positive agglutination

Figure 20.

VIII. Differential tests for distinguishing *B. anthracis* from *B. cereus* (and *B. thuringiensis*)

VIII. 1. Caution

In the normal routine laboratory, if there is reason to believe that an isolate may be *B. anthracis*, culture work beyond what is absolutely essential is not to be encouraged, and such as needs to be done should be done in a Class I exhaust protective cabinet.

However, as one of the oldest known pathogens, *B. anthracis* has, of course, been handled by hundreds of bacteriologists throughout this century and there is no reason why today's bacteriologist, taking sensible precautions, should not be competent to carry out the basic tests necessary to establish the likelihood of an isolate being *B. anthracis* such as Gram and spore stains, and examination for lecithinase production, haemolysis on blood agar, motility and sensitivity to penicillin.

Obviously, culture plates and broths should not be kept longer than is absolutely necessary and should be autoclaved at the earliest opportunity; nor should they be stored on open benches. Incubation and storage should be done with the cultures in suitably marked closed autoclavable containers. Staining procedures should be carried out over a phenolic disinfectant and stains should be washed off into this disinfectant. All materials used should be disposed of in accordance with recognized laboratory safety codes.

In general, if the medical or veterinary history is suggestive of anthrax, thick smears of blood or fluid from pustules will be submitted to the laboratory for Gram and M'Fadyean or Giemsa staining. The presence of Gram-positive rods which, when stained, are seen to be capsulated are essentially considered confirmatory of anthrax. Blood or swabs will also be submitted for culture.

Isolates from other sources having typical *B. cereus*-like colonies but which are non-haemolytic, non-motile and penicillin sensitive should be treated with caution and sent suitably packed with appropriate warnings to the nearest reference laboratory for confirmation. In the United Kingdom, this is the Vaccine Research and Production Laboratory, PHLS–CAMR, Porton Down, Salisbury, Wilts SP4 0JG.

VIII. 2. Background to differential tests for distinguishing B. anthracis *from* B. cereus

From the earliest days of bacteriological history up until the advent of penicillin, bacteriologists were troubled with the problem of differentiating *B. anthracis*, the agent of the much-feared disease, anthrax, from the relatively innocuous *B. cereus* (variously called *B. anthracoides*, *B. anthracis similis* or *B. pseudoanthracis* among other names in former days – Gibson and Gordon, 1974).

Initially, the distinction could only be made by virulence tests in animals. Koch *et al* (1884) inoculated mice by scratching the skin with a needle dipped in the test culture. On the theory that *B. anthracis* 'penetrates into tissues more easily than other (contaminating) organisms' the method is still found being recommended in at least one modern text (Baker and Breach, 1980) for use when putrid material is being tested. The animal acts as a selective agent for re-isolation of *B. anthracis*.

However, Pasteur (1881a, b) had already shown that virulence was a characteristic which could be altered by culture conditions and animal passage and that virulence was not, therefore, an especially good criterion for distinguishing the two species.

The problem clearly had not been fully resolved when Smith *et al* (1946) published their first monograph on *Bacillus* species. They stated that '*B. anthracis* differed from *B. cereus* in being pathogenic and non-motile. Since both are mutable characters, *B. anthracis* has been placed as a variety of *B. cereus*.' Although this was not universally accepted, the term *B. cereus* var. *anthracis* was retained in the later monographs (Smith *et al*, 1952; Gordon *et al*, 1973).

Gordon (1975) admitted that 'the severest criticism of our three reports was levelled at our designating *B. anthracis* and *B. thuringiensis* as varieties of *B. cereus*'. She goes on, however, to defend the stand: 'Because . . . avirulent strains of *B. anthracis* and of *B. thuringiensis* could not be differentiated by our data from strains of *B. cereus*, we believed *B. cereus* to be the stable form of the species. When a strain of *B. cereus* var. *anthracis* or *B. cereus* var. *thuringiensis* loses its virulence its varietal name can be dropped but its species name remains.'

These taxonomic niceties, while of academic importance, still left the clinical laboratory worker with the need to be able to differentiate the potentially highly hazardous anthrax organism from the common, generally non-hazardous *B. cereus* and a variety of differential characteristics apart from animal virulence tests were put forward over the years. The earliest one of significance was what is now referred to as M'Fadyean's reaction (by natural distortion, this has become commonly known now as McFadyean's reaction). M'Fadyean (1903a, b) reported that anthrax bacilli in blood smears (or smears of the oedematous fluid of the throat and lymphatic glands in pigs), when stained with aged methylene blue, could be seen to be surrounded by a characteristic capsule. In fact, it was drawn to his attention (M'Fadyean, 1904) that the observation had actually been made by Weichselbaum in 1892.

Unfortunately, capsule formation does not occur when *B. anthracis* is grown on or in ordinary bacteriological culture media, but re-isolation of capsulated strains from inoculated laboratory animals made a useful diagnostic adjunct to animal pathogenicity tests, especially since, in a proportion of such tests, *B. cereus*, *B. mycoides* and *B. thuringiensis* will prove virulent (Nordberg, 1953; Burdon, 1956; Brown *et al*, 1958). The latter will not, however, be encapsulated upon re-isolation.

Formation of the glutamyl polypeptide capsule was subsequently found to be inducible *in vitro* if the strains were grown in the presence of excess CO_2 on agars containing serum (Sterne, 1937), bicarbonate (Thorne *et al*, 1952), activated charcoal or bicarbonate plus serum or activated charcoal alone (Meynell and Meynell, 1964) or milk (Weaver *et al*, 1970). Meynell and Meynell (1964) attributed the action of the serum or charcoal to absorbing out an inhibitor (thought to be a fatty acid) to capsule formation.

Capsulation on these special media was reflected in the mucoid appearance of the colonies as well as being visible by microscopic observation of suitably stained smears. Meynell and Meynell noted excellent capsule formation on agar containing bicarbonate and serum at pH6.8–7.4 in the presence of 5% CO_2 – conditions, they pointed out, analagous to those found in the mammalian body. Williams and colleagues found growth on bicarbonate agar an effective way to differentiate *B. cereus* and *B. anthracis* (Dr R.P. Williams, personal communication).

Lack of motility was noted at an early stage as being a useful differential characteristic; generally *B. anthracis* is non-motile and *B. cereus* is motile. However, motile strains of *B. anthracis* have been reported (as reviewed by Brown and Cherry, 1955) as have non-motile *B. cereus* strains.

Sensitivity to penicillin G is a useful diagnostic test. We have been notified of the recent isolation of a penicillin-resistant strain of *B. anthracis* (Mr J.A. Carman, personal communication) but no other such reports were found and this is clearly very unusual. Penicillin-sensitive strains of *B. cereus* are occasionally encountered.

Lysis by gamma phage has nowadays become accepted as a highly specific differential test for *B. anthracis* and an isolate with all the characteristics of non-motile, non-haemolytic (or poorly haemolytic) *B. cereus* which is penicillin-sensitive and lysed by gamma phage is essentially, by definition, *B. anthracis*. In a manner not infrequently encountered in science, this phage was discovered 'by chance' and obtained through a procedure whose mechanism of action was not known (Brown and Cherry, 1955). The phage test differs from methods related to demonstration of capsulation in that it lyses virulent and avirulent strains and does not, therefore, only detect virulent strains. On the other hand, Buck *et al* (1963) found that 15% of 264 separate isolates of *B. anthracis* were not lysed by this phage.

Other tests have been proposed for distinguishing *B. anthracis* from *B. cereus* such as comparative reduction of methylene blue, requirement for thiamine in an ammonium salt based defined medium (Proom and Knight, 1955), relative growth at 45°C, salicin fermentation, tyrosine decomposition, the Ascoli capsular precipitin test, mouse lethal tests and tissue culture tests.

Table 6 lists the majority of the proposed differential tests with the generally expected results and the actual results obtained in a study by Brown *et al* (1958) using 137 strains of *B. anthracis*, 115 strains of *B. cereus* and 70 strains of *B. cereus* var. *mycoides* obtained from laboratories across the world. It is of interest that, in addition to the tests referred to above, these authors found that the majority of the strains of *B. anthracis* tested were LV negative.

The results that are listed as expected for *B. cereus* would also be those anticipated for *B. thuringiensis* which, apart from its ability to form its parasporal crystal bodies (see III. 3 above), is indistinguishable from *B. cereus*.

Table 6 Proposed tests to differentiate *B. anthracis* from other members
findings of Brown *et al*

			B. anthracis
		Generally	Brown *et al* (1958) No. +/Total tested
* 1. Motile		–	0/122
2. 'Cotton wool' growth in broth culture		+/–	78/122
* 3. Haemolysis on 5% blood agar		–	45/122
4. LV reaction		–/+	19/89
* 5. Sensitive to pencillin (10 units)		+	n
* 6. Susceptible to gamma phage		+	122/122
7. 'Inverted pine tree' growth in gelatin stab		–/+	21/122
* 8. Capsule formation – fresh animal isolates		+	118/131
9. Capsule formation on bicarbonate medium		+	n
10. Reduction of methylene blue (48 hours)		–	0/122
11. Pathogenicity (0.2 ml 18 h broth culture subcutaneously)	mice	+	107/120
	guinea pigs	+	41/47
	rabbits	+	34/42
**12. Lethal toxin (0.5 ml cell-free culture filtrate intravenously in mice)		–	n
†13. Requiring thiamine		+	n
§14. Growth at 45°C		–	n
‡15. Tyrosine decomposition		–	n

*In general, these together with colony tenacity and Gram stain appearance constitute the basic and most practical set of tests for differentiating *B. anthracis* from other members of the *B. cereus* group.

of the *B. cereus* group. Expected results in general and the (1958).

	B. cereus/*B. thuringiensis*		*B. cereus* var. *mycoides*	
	Generally	Brown *et al* (1958) No. +/Total tested	Generally	Brown *et al* (1958) No. +/Total tested
1.	+ (sometimes feeble)	115/115	Slight/–	30/38 (all slight)
2.	–	16/115	–	4/38
3.	+	89/115	+	38/38
4.	+	89/97	+	15/15
5.	–	n	–	n
6.	–	0/115	–	0/38
7.	–/+	41/115	–/+	13/38
8.	–	26/115	–	3/38
9.	–	n	–	n
10.	–/+	36/115	–/+	7/70
11. mice	–/+	26/63	–/+	3/27
guinea pigs	–/+	7/26	–/+	3/27
rabbits	–	0/24	–	0/27
12.	+	n	n	n
13.	–	n	n	n
14.	+	n	–	n
15.	+	n	+/–	n

n = not done, not recorded, not known or not applicable or only a few strains tested.
**Bonventre and Johnson (1970). †Proom and Knight (1955). §Burdon (1956).
‡Gordon *et al* (1973).

21

A

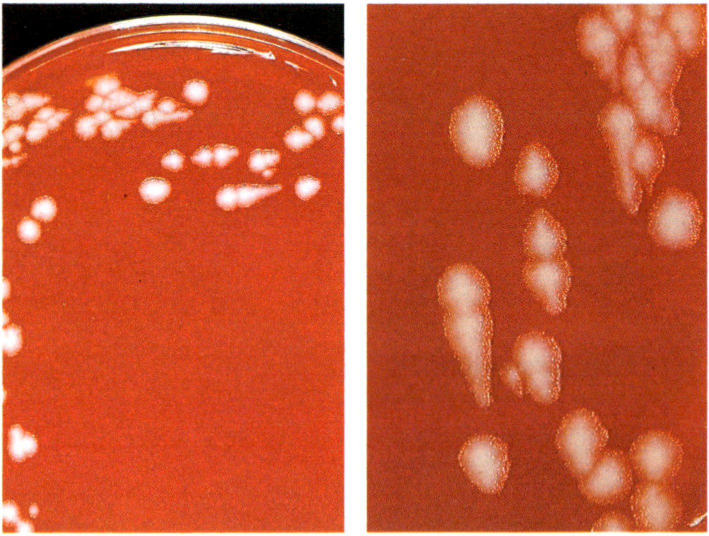

B

A, × 1. **B**, × 6. Highlighting 'spikes'.

Figure 21 ***B. anthracis*** colonies (*B. anthracis* A19432/80; 37°C for 24 hours).

VIII. 3. Colonial characteristics (**Figure 21**) *'Medusa Head'*

There is very little to distinguish *B. anthracis* from *B. cereus* on a first glance at their colonial appearance. However, typical colonies of *B. anthracis* are very viscid in consistency and show a marked tenacity ('stringy' and 'tacky', resembling drying glue) on drawing a loop across them. The strings of growth produced by drawing a loop across a colony can be made to stand up at right angles to the agar surface without support (a picture of this is shown in Figure 2 of the chapter by Feeley and Patton, 1980). This tenacity is attributable to the organisms' growth in long strings of cells, as can be seen with a Gram stain (**Figure 22**).

This form of growth, in turn, is probably responsible for the 'spiking' visible from some of the colonies (**Figure 21**). Spiking or tailing, often along the inoculation line, is regarded by those familiar with *B. anthracis* as being associated with virulence. The colonies of avirulent strains tend to be essentially circular and lack these outgrowths (Mr. B.J. Capel, personal communication). The spikes may appear as comma-shaped outgrowths from the colony as seen in Figure 1 of the chapter of Feeley and Patton (1980).

As with most of the tests proposed for differentiating *B. anthracis* from *B. cereus*, significant tenacity should only be regarded as a signal to be alert and not diagnostically reliable in its own right. For example, *B. anthracis* strains NCTC 2620 and 7753 exhibited no tenacity and were indistinguishable in appearance on Gram staining from what would normally be expected with *B. cereus*. Whether, in the case of these strains, this is due to age and storage is, of course, uncertain, but equally, tenacity may be encountered on occasion with colonies of *B. cereus*. This generally appears to be a 'softer' tackiness, but from time to time, strains of *B. cereus* are found with a highly stringy consistency such that growth can be teased with a loop or wire into standing perpendicular to the agar surface. These are associated with long strings of cells indistinguishable upon Gram staining from tenacious *B. anthracis* (see VIII. 5 and **Figure 22**).

Colonies of *B. anthracis* rarely show the greenish tinge frequently (but not invariably) seen in colonies of *B. cereus*; this is undoubtedly related to haemolysis (see under **Figure 33**).

'Medusa Head'

Reference is frequently made in texts to the 'Medusa Head' appearance of colonies of *B. anthracis* – i.e. highly tangled outgrowths from the edges of the colonies. This is, in fact, a common feature of colonies of the *B. cereus* group as a whole and can be seen with colonies of many strains of *B. cereus*, *B. thuringiensis* and *B. cereus* var. *mycoides* as well as *B. anthracis*.

Candeli *et al* (1979) also noted this appearance in colonies of some strains of *B. megaterium*. Thus, as far as differentiation of *B. anthracis* from its close relatives is concerned, this is not a test of value.

The procedure given in some texts involves placing a cover-slip on top of young colonies on an agar surface and examining the colony edges under a × 10 objective. By today's laboratory safety criteria, it would probably be unacceptable to do this with conventional Petri dishes; the culture would have to be grown in some suitable container from which the lid need not be removed for examination under the microscope.

Photographs of 'Medusa Head' colonies can be seen in Figures 1 and 3 of the paper of Candeli *et al* (1979) and Figure II. 31, page 79, of Stratford's *Atlas of Medical Microbiology* (Stratford, 1977).

VIII. 4. Haemolysis **(Figures 23, 24)**

In general, colonies of *B. anthracis* appear non-haemolytic or, at most, weakly haemolytic on conventional horse or sheep blood agar. Although great variability is encountered with *B. cereus* and, in our experience, very weakly haemolytic strains are not unusual, strong haemolysis is more frequent than not around colonies of *B. cereus*.

The contrast can be seen in **Figures 23** and **24** between the haemolysis produced by *B. cereus* strain A93781/81 (from a pus swab) and its absence in the case of *B. anthracis* strain A19432/80 (from a necrotic lesion on the forearm of a hide-sorter. The lesion had a vesiculated spreading edge – the 'malignant pustule' of anthrax. This strain was kindly supplied by Dr S. Miller, South African Institute for Medical Research.)

As, again, with virtually all the tests for differentiating *B. anthracis* and *B. cereus* when taken on their own, haemolysis (or lack of it) taken alone can only provide a guide and is not diagnostically reliable in the absolute sense. Brown *et al* (1958), discussing the results shown in **Table 6**, commented that they felt no sharp distinction could be made between the species on the basis of haemolysis.

VIII. 5. Gram stain **(Figure 22)**

Reference should be made to section VIII. 3 where it was noted that typical colonies of *B. anthracis* show a marked tenacity which is associated with a Gram stain appearance such as seen in **Figure 22A**. The bacterial cells are seen in long, seemingly endless strings which often appear greatly entangled. Less dramatic strings are seen in **Figure 2** (p. 26).

The bacteria in Gram stained smears from colonies of *B. cereus* appear singly, as in **Figure 22B**, or in relatively short chains – i.e. the ends are visible. Very occasionally, highly tenacious colonies of *B. cereus* are encountered which give a Gram stain appearance resembling **Figure 22A** and, likewise, non-tenacious strains of *B. anthracis* may be found which give a Gram stain appearance nearer to **22B** than to **22A**.

22

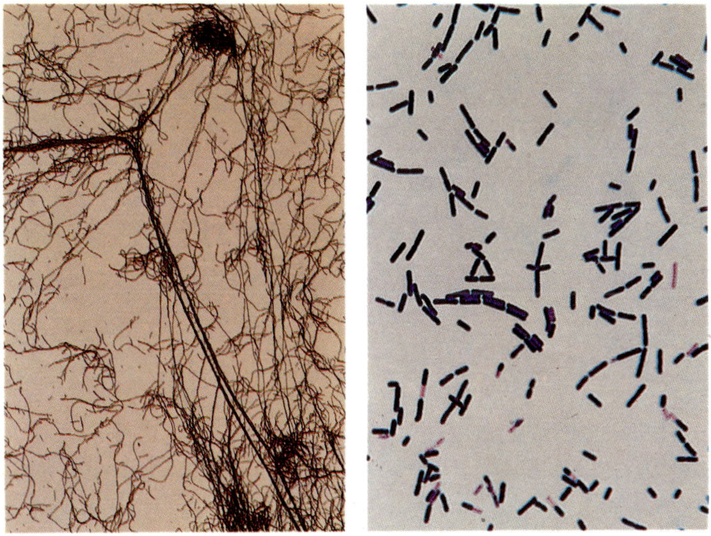

A B

A *B. anthracis* NCTC 5444, ×60. **B** *B. cereus* F4433/73, ×600.
Figure 22 Gram stain (from colonies on blood agar after 24 hours at 37°C).

VIII. 6. Motility
Motility tests were covered in section III. 4.

B. anthracis is generally regarded as being non-motile while *B. cereus* is normally motile.

One report of the existence of motile strains of *B. anthracis* was found (Brown and Cherry, 1955). On the other hand, non-motile, or only very sluggishly motile strains of *B. cereus*, though not frequent, are encountered periodically.

***VIII. 7. Sensitivity to penicillin* (Figure 23)**
Sensitivity to penicillin was originally tested on tryptose or trypticase soy agar containing penicillin G at a final concentration of 5–10 units/ml. A loopful of a light saline suspension of the organism under test was spread across this. Penicillin G discs are more appropriate today.

23

A

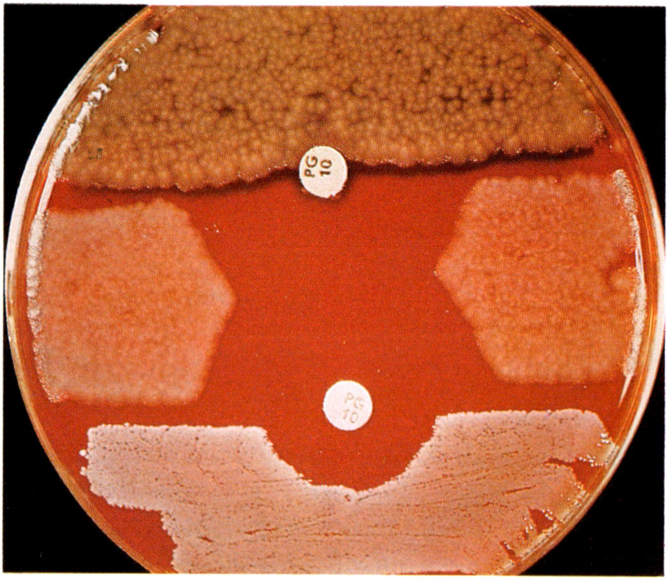

Figure 23 **Sensitivity to penicillin**. Modified Stokes' test.

Top: *B. cereus* A93781/81. Resistant. (Note also the zone of haemolysis along the edge of the line of growth.)

Middle: *B. anthracis* A19432/80. Sensitive. (Note also the absence of haemolysis.)

Bottom: *Staphylococcus aureus* NCTC 6571 (The 'Oxford Staph'). Sensitive control.

(a) Procedures using sensitivity discs

A standard sensitivity test procedure such as the Stokes' method is probably the best approach (**Figure 23A**) although the discs can be used directly (**Figure 23B**) if a suitable control is included on a separate plate to check the efficacy of the discs.

Where for most bacteria, slightly turbid peptone water suspensions (equivalent to McFarland's or Brown's barium sulphate opacity standards 0.5) are recommended for preparation of suitable lawns, for *B. anthracis* and *B. cereus* which grow rapidly and produce large colonies, the peptone water suspensions should show virtually no turbidity at all.

23

B

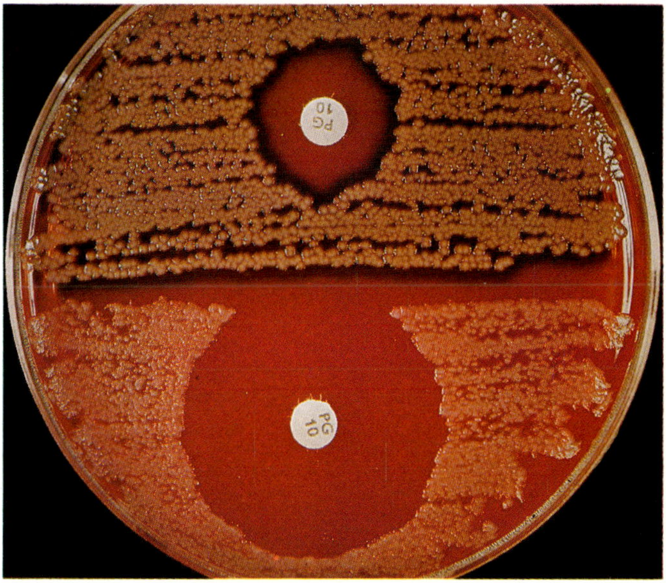

Figure 23 Sensitivity to penicillin.
Top: *B. cereus* F4433/73. Moderately resistant. (Note also the haemolysis.)
Bottom: *B. anthracis* NCTC 7753. Sensitive. (Note the absence of haemolysis.)

Prepare the peptone water suspensions by inoculation with a straight wire touched to 3–5 colonies of the organisms under test.

Dip a swab into the peptone water suspensions, remove excess fluid by squeezing and rotating the swab against the side of the container above the suspension, and streak this across the appropriate region of the test agar plate. Place the penicillin G discs in the appropriate positions and incubate at 37°C overnight. *B. anthracis* is sensitive; *B. cereus*, is with the occasional exception, resistant or moderately resistant.

The incubation temperature is important; according to Knox and Collard (1952), sensitivity of *B. cereus* to pencillin is greatly enhanced at 41°C, at which temperature, synthesis of the inducible enzyme, penicillinase, is suppressed.

23

C

Figure 23 Sensitivity to pencillin. 'String of Pearls' test. ×3400 (oil immersion objective). (Photograph kindly supplied by Mr R. Charlton, Bulawayo Group Laboratory, Zimbabwe.)

(b) 'String of Pearls' test

This test was first described by Jensen and Kleemeyer (1953). They observed that *B. anthracis* produced large round cellular forms in chains ('strings of pearls') when grown on tryptose agar containing 0.05–0.5 units/ml of penicillin for 3–6 hours. The phenomenon reflects the altered morphology that *B. anthracis* cells undergo in the presence of relatively low levels of penicillin; the bacilli swell and appear as chains of spheres.

During a recent very large outbreak of anthrax in Zimbabwe (more than 6000 human cases – Turner, 1980), in the absence of other readily available diagnostic tests, this test was found to be very useful in confirmatory diagnoses (Mr R. Charlton, personal communication).

Buck *et al* (1963), examining 264 isolates of *B. anthracis* for 'strings of pearls' production, noted that a number of the strains did not grow – thereby implying a disadvantage; however, the recorded level of penicillin in the tryptose agar they used was 5–10 units/ml.

Bailie and Stowe (1977) report that 'strings of pearls' can be seen in the vicinity of a 10 unit penicillin G disc on a Mueller–Hinton plate inoculated with *B. anthracis*.

Procedure (Charlton, 1980)

(i) Prepare a solution of sodium benzyl penicillin in sterile distilled water to contain 50 units/ml. Add 1 ml of this to 100 ml of melted blood agar base (without blood) and pour into four 9 cm Petri dishes (i.e. 25 ml/dish). Allow to set.

(ii) Using a scalpel blade, cut a block approximately 1.5 cm square from the penicillin agar plate and place it on a microscope slide. Put the slide in a Petri dish together with a small piece of moistened cotton wool (to keep the small agar block from drying out).

(iii) Make a line with a suitable marker along the length of a clean cover slip and about 5 mm from one edge.

(iv) Touch a loop to the edge of a young colony of the suspect culture and streak it along the centre of the agar block on the slide.

(v) Place the cover slip so that the line made in (iii) is *face down* alongside the streak of culture. The line then acts as focusing and location guide.

(vi) Put the lid on the Petri dish and place in a 37°C incubator.

(vii) After 2 hours, place the slide on a microscope stage. Focus on the line with the ×10 objective and bring the high dry or oil immersion lens into use to look for the 'strings of pearls'.

(viii) Note that vigorously growing cells are necessary for reliable results – hence the term 'young' colony in (iv) above.

24

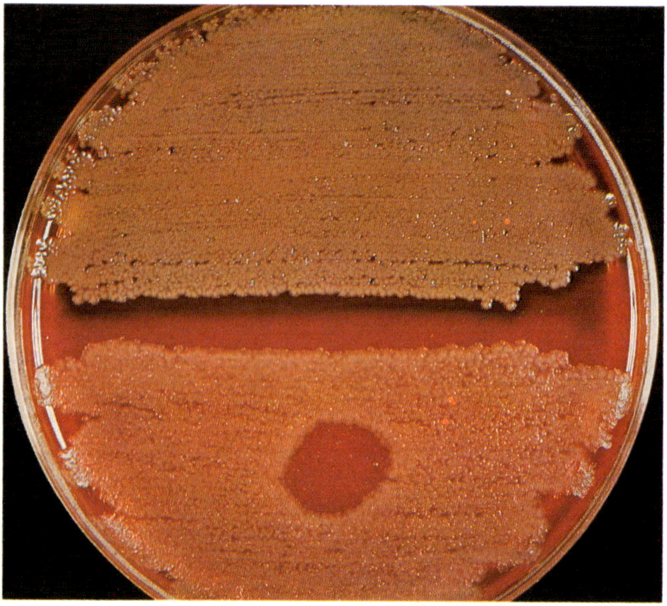

Figure 24 Gamma phage (24 hours at 37°C).
Top: *B. cereus* F4433/73 (note also the haemolysis).
Below: *B. anthracis* A19432/80 (note also the absence of haemolysis).

VIII. 8. Gamma phage (Figure 24)

(a) Phage origin

McCloy (1951a) isolated a bacteriophage from an atypical strain of *B. cereus* known as strain W; the phage was also termed phage W and it attacked strains of *B. anthracis* (but some only feebly and not if they were in the mucoid state), the parent *B. cereus* strain W and 2 of 56 other strains of *B. cereus* but none of 18 other species of *Bacillus*. She subsequently concluded (McCloy, 1951b) that W phage actually consisted of a temperate phage Wß with which *B. cereus* strain W was lysogenic and a virulent mutant of Wß, named W$\propto$, which could lyse strain W.

‘By chance’, Brown and Cherry (1955) observed that a third (variant) phage, which they called γ-phage, could be obtained from the original W bacteriophage. This had the ability to lyse smooth forms of *B. anthracis* and did not lyse any of the *B. cereus* strains tested.

McCloy herself (1958) isolated another mutant of the temperate phage Wß and she also referred to this as phage γ. However, this appears to be a separate entity from the γ-phage of Brown and Cherry (1955) with different properties. The parent phage used in the test described below was that of Brown and Cherry.

In a test on 74 unrelated *B. anthracis* strains (Buck *et al*, 1963), 63 (85%) were lysed by the γ-phage. Of 64 strains of other species of *Bacillus* that were examined, just a single strain of *B. megaterium* was lysed. Brown *et al* (1958) found that γ-phage lysed all of 122 strains of *B. anthracis* they tested, none of 115 strains of *B. cereus* and 8 of 70 strains of *B. cereus* var. *mycoides.*

(b) Procedure

(i) Inoculate a blood agar plate (or other suitable agar such as brain–heart infusion or heart-infusion agar) evenly over its entire surface with a loopful of the test culture.

(ii) If necessary, allow the plate to dry for a few minutes.

(iii) Place a loopful or small drop of phage suspension in the centre of the plate (do not spread) and incubate at 37°C.

(iv) The phage inhibition (or its absence) should be readily apparent at 6–8 hours of incubation although it can be read after overnight incubation – as was the case in **Figure 24**.

(v) A control culture of *B. anthracis* should, ideally, be included also.

(c) Propagation of the gamma phage

A culture of *B. anthracis* which is susceptible to gamma phage should be checked for purity. Brown and Cherry (1955) used either strain Ohio or strain Davis; the latter is available as ATCC 11949. We have successfully used NCTC 7752 which was chosen at random.

Nutrient broth is inoculated with a loopful of culture from the purity plate and incubated with gentle shaking at 37°C until slightly turbid (usually about 3–3½ hours in cultures of 10–100 ml). At this point, 0.3–0.5 ml of phage suspension/10 ml of culture is added and incubation continued for approximately 6 hours. Successful phage growth is indicated by the return of the culture to complete clarity.

The culture is then filtered through a Seitz or membrane (0.22 μm) filter. The filtrate contains the phage which may be stored in the refrigerator (it should not be frozen).

The phage titre may be determined by making 10-fold dilutions of the filtrate down to 1 in 10000 and putting 0.02 ml drops of the neat filtrate and the dilutions on to a plate seeded with the propagation strain. The plate is incubated at 37°C overnight and the titre read as the highest dilution giving nearly confluent plaques.

VIII. 9. Capsulation

(a) General information

The capsule of *B. anthracis* (in contrast to the polysaccharide capsules of most other bacteria which produce capsules) is a polypeptide – poly-D-glutamic acid. The appropriate stains for this have been found to be polychrome methylene blue (M'Fadyean reaction stain), Giemsa's stain or 0.1% toluidine blue in 1% alcohol. Immunofluorescence has also been used (Cherry and Freeman, 1959).

It should be noted that other *Bacillus* species, notably *B. subtilis* and *B. megaterium* also produce polypeptide capsules visible by staining and positive in immunofluorescence using the antiserum prepared against capsulated *B. anthracis* (Cherry and Freeman, 1959; Weaver *et al*, 1970). However, in the usual situation under which examination for encapsulated Gram-positive bacilli is going to be made – that is, on specimens of blood or tissues or from lesions of animals which have died or of humans with appropriate case histories – these other *Bacillus* species would be unlikely to be encountered.

B. anthracis organisms will not possess capsules unless they have grown under conditions conducive to capsule formation. Examination for capsules may thus be made on:

(i) Specimens such as serosanguinous discharge from the vesicular pustules of cutaneous anthrax, or blood (or, in the case of pigs*, oedematous fluid from the throat region) from animals suspected of having had anthrax,

(ii) blood or tissue impression smears from laboratory animals inoculated for pathogenicity tests and re-isolation purposes, and

(iii) cultures grown on bicarbonate agar with serum in the presence of carbon dioxide (see VIII. 9c).

*In most animal species, at the time of death from anthrax, the bacilli will have become generalized and be present in the blood stream in large numbers. In the pig, however, the blood may still be free of, or carrying only very low numbers of the organism; characteristically, the throat becomes swollen and oedematous prior to death.

25

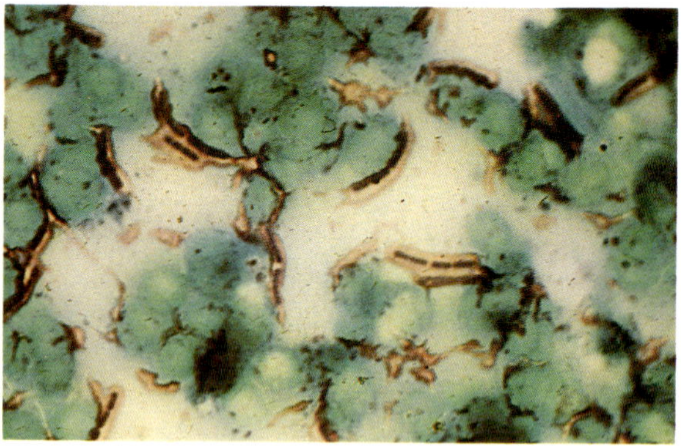

***Figure 25 Encapsulated *B. anthracis*.** M'Fadyean stain. Bovine blood smear. × 500.

(b) Staining procedures

Immerse air-dried films from (a) (i), (ii) or (iii) above for 3 minutes in absolute methanol or 1:1000 mercuric chloride to fix the films and stain by one of the methods below:

(i) *M'Fadyean's reaction* **(Figure 25)**

- Cover the smear with polychrome methylene blue for 30 seconds.
- Wash with water (into disinfectant†), blot (discarding blotting paper into an autoclave bin) and allow to dry.
- Examine under oil immersion.

The bacteria stain blue and the capsule pink (**Figure 25**). Some texts recommend heat-fixing the smears, in which case, the capsule breaks down and is seen as an amorphous, granular, reddish-purple material surrounding the blue bacterial cells. A picture of this can be seen in Stratford's *Atlas of Medical Microbiology* (Stratford, 1977), Figure III. 32, page 74.

*Crown Copyright. Photograph kindly supplied by courtesy of the Library, Central Veterinary Laboratory, Weybridge, Surrey, England.
†See footnote on page 79.

(ii) *Giemsa stain*

- Immerse the fixed film in freshly diluted (1 part stock: 10 parts 0.01 M phosphate buffer, pH 7.0) Giemsa stain and leave for 30 minutes.
- Wash the slide by immersing it in plain buffer for 30 seconds.
- Blot and allow to dry.
- Examine under oil immersion.

The blue to purple bacilli are surrounded by a pink to red capsule.

(iii) *0.1% toluidine blue in 1% ethanol*

- Cover the fixed film with this stain and leave for 5 minutes.
- Wash (into disinfectant*).
- Blot and allow to dry.
- Examine under oil immersion.

The capsule is seen as a light blue zone around the dark blue bacterial cell.

(iv) *Fluorescent antibody detection of capsules*

A number of texts are to be found suggesting immunofluorescence as a suitable procedure for detecting encapsulated *B. anthracis*. This largely stems from the work of Cherry and Freeman (1959). The method requires antibody raised to highly encapsulated cells.

Despite its proposed use in various texts, the procedure does not recommend itself for the following reasons: (A) specific antisera to encapsulated *B. anthracis* is not readily available; (B) it is fairly non-specific and other *Bacillus* species may give positive results; (C) attempts by Cherry and Freeman to absorb out the non-specific antibodies simply removed its ability to give fluorescence with *B. anthracis*; (D) the procedures, especially the indirect method, are quite lengthy and cumbersome for the busy routine laboratory.

*The fixing process will probably not have killed spores present in the smear. Precautions such as washing the stains off into disinfectant are wise. Phenolic disinfectants such as 3% Lysol or 2% Hycolin and 5% phenol itself have some sporicidal activity, but for double safety, the disinfectant in its container should also be autoclaved on completion of work. 5% potassium permanganate solution is effectively sporicidal within 15–30 minutes and may be used as an alternative to the phenolic disinfectants.

In view of our feeling that immunofluorescence is of no significant value to the routine laboratory in the differential identification of *B. anthracis*, detailed coverage of the methods involved was deemed out of place in this text. If the anti-capsulated cell serum is available, standard direct or indirect immunofluorescent 'staining' methods can be used. The procedures are well described in numerous other texts.

(v) *India ink*

One reference was found (Cherry and Freeman, 1959) to the use of the India ink method (as generally applied to observing polysaccharide capsules formed by a number of bacterial species) for detecting encapsulated *B. anthracis*. This would only be applicable to cells from pure cultures from appropriate culture plates (see below) as opposed to those in tissues of infected animals.

(c) Encapsulation on artificial media (Meynell and Meynell, 1964)

(i) Inoculate sodium bicarbonate plus serum plates to obtain isolated colonies.

(ii) Incubate the plates in a carbon dioxide incubator with 5–10% carbon dioxide or in a candle jar or in an anaerobic jar without catalyst containing a carbon dioxide generator at 37°C overnight.

(iii) Good capsulation is obtained with carbon dioxide levels of 5–40%.

VIII. 10. Pathogenicity tests and re-isolation

(a) General information

Animal tests are essential to definitive identification of *B. anthracis*. However, they may be necessary on occasion to establish virulence. Capsulation, while essential to virulence, presumably through preventing uptake of the organisms by host macrophages, is not necessarily indicatory of virulence and avirulent encapsulated strains occur (Thorne *et al*, 1952; Meynell and Meynell, 1964).

The actual cause of death in lethal anthrax has not been fully established. Virulence appears to depend on both capsule formation and ability to produce a toxin. The precise nature and site of action of this toxin has yet to be determined. It is clear, however, that observation of a capsule, while being a highly significant differential characteristic for distinguishing *B. anthracis* from *B. cereus*, is not adequate as an indicator of virulence.

(b) Safety precautions

Pathogenicity tests and re-isolation of *B. anthracis* in laboratory animals must be carried out with great caution to avoid distribution of the sporing organisms.

Inoculations should be carried out in a Class I exhaust protective cabinet and the inoculated animals should be kept in housing which (i) is separate from that of other animals and (ii) allows easy and effective sterilization of litter, excreta, cages and, after these have been removed, of the room, racks etc.

Postmortems should be carried out on aluminium foil, in turn, in metal trays (and, of course, in a Class I exhaust protective cabinet); at the end of the postmortem, the foil can be wrapped over the carcass and this can be put into a suitable bag or container for autoclaving followed by incineration. The tray and instruments used in the postmortem should also be autoclaved.

Disposable gloves should be worn during all procedures involving handling of the infected animals or their carcasses, litter, excreta and cages.

All working surfaces should be washed with 5% phenol or 5% hypochlorite solutions.

It is recommended that this work be left to a suitably equipped reference laboratory where possible.

(c) Inoculation and examination

Considerable variation is found in text-books and papers with respect to dose, route of inoculation, site of inoculation, susceptibility and suitability of different laboratory animal species and time to death. No standardized test has been laid down.

Furthermore, it is clear from the literature that there is a significant cross-over between results obtained with less virulent and avirulent strains of *B. anthracis* and virulent strains of *B. cereus*. Brown *et al* (1958) review this topic well.

Since these authors found a high proportion (25 of 63–40%) of strains of *B. cereus* killed mice with pathology typical of anthrax while a lower proportion (7 of the 25) were correspondingly virulent for guinea pigs, they proposed that mice should be used for the screening of pathogenicity with the guinea pig being used for confirmatory tests. Their doses however (0.2 ml of a 6–18 hour heart-infusion broth culture incubated at 37°C), would have been considerably larger than that used by most workers.

With respect to route of inoculation, most texts recommend 'inoculation of 0.2–0.5 ml subcutaneously in mice or guinea pigs or intraperitoneally in mice or intramuscularly in guinea pigs'. At least for the operator working alone, intraperitoneal injection in a mouse is a far simpler procedure than subcutaneous injection.

The following procedure is an attempt at a sensible amalgam of the many and various methods recorded in the literature.

(i) A slightly turbid saline suspension of the organism (about 0.5 on a McFarland's or Brown's opacity standard) should be made from an overnight agar culture. This should be diluted 1:10.

(ii) 0.25 ml of this suspension should be injected intraperitoneally into at least duplicate mice or subcutaneously or intramuscularly (thigh region) into guinea pigs.

(iii) Death of the animals should result in 1–3 days.

(iv) On postmortem examination, gelatinous oedema at the site of inoculation, especially with the subcutaneous and intramuscular routes, may be apparent.

(v) Smears from the oedematous fluid, heart blood and impression smears from the cut surface of the spleen should be examined for the presence of the encapsulated bacteria (see VIII. 9b). These specimens should also be plated on blood agar for culture (37°C for 24 hours). (It is just a practical observation without obvious reason that the capsules may seem most clear to the less experienced in spleen impression smears.)

(vi) If the animals have not died within 72 hours, they should be held for 10 days and possibly the test repeated using fresh animals with a 10-fold higher dose before concluding that the strain under test is avirulent.

VIII. 11. Reduction of methylene blue (Burdon, 1956)

(i) A methylene blue agar tube is inoculated from a colony of the culture being tested with a single stab to the bottom of the tube using a straight wire.

(ii) The tube is incubated at 37°C overnight and examined for growth and loss of colour in comparison with an uninoculated control.

(iii) *B. anthracis* generally produces only slight or no reduction of methylene blue while *B. cereus* more frequently causes a marked reduction within 24 hours (Brown *et al*, 1958 – see **Table 6**).

VIII. 12. Growth at 45°C (Burdon, 1956)

(i) Tryptose agar slants are pre-warmed in a water bath to 45°C.

(ii) The slants are inoculated with a single streak using a straight wire and are returned at once to the water bath.

(iii) According to Burdon (1956 – no other reports known to us), *B. anthracis* shows little or no growth at 24–48 hours while *B. cereus* shows good growth.

VIII. 13. Requirement for thiamine (Proom and Knight, 1955)
Since this test requires preparation of a defined medium (see 'Defined complex amino acid medium' in the Media and Reagents section, page 257) with or without thiamine, it is included here more for the sake of completeness than as a practical test for the normal clinical laboratory.

(i) The inoculum is picked with a small loop carefully from the top of a colony of the *Bacillus* on a blood agar plate (to avoid carrying over nutrient material from the agar plate). It is transferred to tubes of the complex amino acid medium with and without thiamine.

(ii) The tubes are incubated for 48 hours at 37°C and observed for growth at 24 and 48 hours.

(iii) *B. cereus* should grow readily in both tubes; according to Proom and Knight (1955), *B. anthracis* should only grow in the tube with thiamine. If growth is found in both tubes, then a second tube without thiamine should be inoculated from the first tube without thiamine using a straight wire. Growth is again looked for after further incubation at 37°C for 48 hours. The purpose of this second passage is to ensure that carry-over of essential nutrients from the blood agar plate did not account for the growth in the first tube.

As described in their paper of 1955, Proom and Knight's test was done with 6 strains of *B. anthracis*. Three strains of *B. thuringiensis* did not require thiamine. *B. cereus* was not included in the 1955 study; its ability to grow in the absence of thiamine is to be inferred from earlier work (Knight and Proom, 1950) using 13 strains of *B. cereus*. No other references to testing thiamine requirements were seen.

VIII. 14. Decomposition of tyrosine

This test has been described in section VI. 3. The readings given by Gordon *et al* (1973) indicated that it can be used as a differential test for *B. anthracis* and *B. cereus*.

B. anthracis should show no zone of clearing in the tyrosine agar; *B. cereus* gives a positive reaction.

VIII. 15. Ascoli precipitin test (Ascoli, 1911)

This is a diagnostic test for the presence of *B. anthracis* antigens in the tissues of infected animals. Its primary value is in situations where rapid retrospective evidence of animal infection is required – for example, in detecting whether hides have derived from animals that were infected with anthrax.

The test is quite frequently proposed in text-books but, as with immunofluorescence (see VIII. 9b. iv), it requires the appropriate antiserum. This is not readily available and the test is, therefore, not one that can be carried out in most routine clinical laboratories.

Also as with immunofluorescence, the test is not rigorously specific for *B. anthracis*. The antigens involved are probably common to several of the *Bacillus* species, but, on the other hand, it is unlikely that other *Bacillus* species would have proliferated within and throughout an animal as *B. anthracis* does and which deposits extensive precipitating antigens in the tissues.

Procedure

(i) 2 g of the tissue under test are boiled for 5 minutes in 5 ml of saline containing 1:100 (final concentration) acetic acid.

(ii) After cooling, the fluid should be filtered through filter paper. The antigen, if present, will be in the filtrate.

(iii) Using a capillary tube as in the Lancefield test for streptococcal grouping or a small (4 mm cross-section) test tube, the antigen is layered on top of the antiserum. Development within 15 minutes of a white ring of precipitate at the interface of the two fluids is a positive reaction.

(iv) An attempt to confirm the positive result (and indeed to check negative ones) by culture and other tests described in this section VIII should be made.

Preparation of antiserum for Ascoli test

The following two methods were kindly supplied by Dr J. Bergmann, Veterinary Diagnostic Centre, Windhoek, South West Africa/ Namibia.

Method 1 involves inoculation of live virulent bacteria in pre-immunized rabbits and method 2 uses formalin-killed virulent cells in otherwise unimmunized rabbits. Method 1 produces the antiserum with the higher titre but suffers the disadvantages of (a) requiring use of live virulent *B. anthracis* and (b) a greater casualty rate among the inoculated animals. Method 2 is safer from the handler's point of view, but is more spasmodic than method 1 in its effectiveness. Between 30 and 50% of inoculated animals will produce a usable antiserum.

(A) Method 1

(i) Injections 1 and 2 – on days 1 and 14, inject rabbits subcutaneously with 0.5 ml of the live Sterne strain anthrax vaccine (see Appendix 1, **Figure 154**).

(ii) For subsequent inoculations, the inoculum is prepared by growing a mixture of as many strains of *B. anthracis* as possible overnight (37°C) on an agar slope in a standard 15 mm cross-section test-tube. The growth is washed off and suspended in 5 ml of physiological saline.

(iii) Injection 3 – on day 28, 0.05 ml of the suspension from (ii) is inoculated into the rabbit subcutaneously.
(iv) Injection 4 – after a 3–5 day interval, a second subcutaneous injection of a suspension prepared as in (ii) is given.
(v) Injections 5, 6, 7 – after further 3–5 day intervals, 0.1 ml volumes of suspensions prepared as in (ii) are administered intravenously.
(vi) Following a resting period of 5–7 days, blood may be taken for assay of serum antibody as in C below. Further intravenous injections may be administered in attempts to raise the antibody titre higher if unsatisfactory at this stage.

(B) Method 2

(i) Agar slopes in standard 15 mm test tubes are inoculated with a mixture of as many strains of *B. anthracis* as possible and incubated overnight at 37°C. The growth is harvested in 0.2% formalized physiological saline, 5 ml/tube. This is held for 2 weeks before use.
(ii) Rabbits are inoculated with 5 consecutive intravenous injections 2–4 days apart and consisting of increasing doses of the formalized suspension from (i) as follows: 0.5 ml, 1 ml, 3 ml, 5 ml and 10 ml.
(iii) One week after the 10 ml injection, blood may be taken for assay of serum antibody as in C (below). If unsatisfactory, a further 10 ml dose should be administered and serum antibody retested after another week.

(C) Serum antibody assay of Ascoli test antiserum

(i) Agar slopes in standard 15 mm test tubes are inoculated with a mixture of as many strains of *B. anthracis* as possible and incubated overnight at 37°C. The growth is suspended in physiological saline (5 ml/tube) and the suspension steamed for 30 minutes. Using a 10% phenol solution, the suspension is brought up to a phenol content of 0.4% final concentration. This is then Seitz or membrane filtered (centrifugation and pre-filters may be necessary).

(ii) A little of the serum under test is put into a serum tube (4 mm cross-section). The phenolized extract from (i) is layered carefully over the serum and a precipitation at the interface is looked for within a maximum of 2 minutes.

A crude titration can be carried out by using serum dilutions.

VIII. 16. Gelatin tests

The test for liquefaction of gelatin is given in section IV. 10. It is frequently stated in texts that gelatin is liquefied only slowly by *B. anthracis* with a characteristic 'inverted fir-tree' pattern being produced around the site of the stab, while *B. cereus* liquefies gelatin rapidly without development of this pattern.

The results obtained by Brown *et al* (1958) (**Table 6**) indicate that this does not provide an especially reliable differential characteristic for these two organisms.

VIII. 17. Salicin fermentation

Salicin fermentation should be tested as an ammonium salt sugar – see section IV. 12. A tube has been included in the bottom right picture in **Figure 18**.

Texts frequently give salicin fermentation as a differential test for *B. anthracis* (negative) and *B. cereus* (positive, sometimes negative – some texts just give *B. cereus* as salicin positive). It is our experience that *B. cereus* is frequently salicin negative (as exemplified in **Figure 18**). The results of Brown *et al* (1958) not only agree with this but also show that *B. anthracis* is by no means invariably salicin negative.

VIII. 18. Peptonization of milk

Burdon (1956) gives peptonization of milk as a differential characteristic for *B. anthracis* (slight, delayed) and *B. cereus* (marked, rapid). In our experience (Parry, unpublished results) wide variations were found among a large number of strains of *B. cereus* grown in milk.

VIII. 19. Lethal toxin

It has been well known for some time that cell-free filtrates of cultures of *B. cereus* in a rich growth broth such as brain–heart infusion broth, are lethal to mice when injected intravenously (0.5 ml into the tail vein). The lethal action is extremely rapid with most strains (usually less than 2 minutes, rarely more than 15 minutes) though Turnbull (1981) found that 12 of 88 (13.7%) strains of *B. cereus* gave negative results.

As reviewed by Bonventre and Johnson (1970), similar preparations of *B. anthracis* do not have this lethal activity and they suggested that it might serve as a useful supplementary differential test to distinguish *B. anthracis* from *B. cereus*.

It has been shown (Turnbull, 1981) that the lethal activity results from the combined effects of two toxic factors produced by *B. cereus*, one of which is the major haemolysin, cereolysin. The negative or weak haemolysin production by *B. anthracis* thus accounts in part for the reported lack of lethal action by analagous *B. anthracis* preparations.

Procedure

(i) Inoculate 10 ml of brain–heart infusion broth (Difco or Gibco) with 0.5 ml of an overnight nutrient broth culture (37°C) and incubate at 37°C on a mechanical shaker fairly vigorously for 6½–7 hours. With very few exceptions, the cultures will grow well under these conditions, and crude pH control by adding 0.2 ml of a 1 N NaOH solution at 4 hours ensures optimal levels of lethal toxin in the final filtrate.

(ii) Centrifuge and filter through a 0.45 μm membrane filter.

(iii) Inject 0.5 ml of the cell-free filtrate into the tail veins of duplicate mice. Return these to their cages and observe for lethal effect within a few minutes (maximum 30 minutes). (We have found that immersion of the tail in water at about 50°C for approximately 30 seconds just prior to injection expands the tail veins and makes injection easier. A 27 gauge needle is advisable.)

Although the number of *B. anthracis* strains we have tested is low, we have confirmed that none of the culture filtrates of 8 culture collection strains of *B. anthracis* and strain A19432/80, prepared as described here, produced observable ill effects when injected intravenously in mice.

VIII. 20. Floccular ('cotton wool') sediment in broth cultures

The frequent tenacity of colonies of *B. anthracis* was attributed in VIII. 3. to the manner in which *B. anthracis* can grow in seemingly endless tangled strings of cells. For the same reason, when this type of growth occurs in broth cultures of *B. anthracis*, these cultures, when shaken up, have the appearance of having pieces of cotton wool floating in them in contrast to the evenly dispersed cloudiness of most bacterial cultures. This flocculent type of growth can even occur in cultures grown on a shaker.

On the rare occasion, tenacious strains of *B. cereus* are encountered which will also grow in this way.

VIII. 21. Selective media

As listed in Table 33.1, page 450 of the text-book of Cruickshank *et al* (1975), a number of selective media for isolation of *B. anthracis* have been proposed. Mostly these are designed to inhibit unrelated bacteria. Claims that *B. anthracis* will grow while *B. cereus* is inhibited were made for the PLET medium of Knisely (1966), the PEA medium referred to in Table 4.2–7 of the text-book of Washington (1981) and a blood agar medium containing 100 mg/litre of trimethoprim as the active ingredient (Oppenheim and Koornhof, 1980).

With PLET medium, inhibition of *B. cereus* growth was attributed to the combined effect of EDTA and thallous acetate in depriving the *B. cereus* of an essential cation. It was proposed that *B. anthracis*, but not *B. cereus*, could utilize thallium in place of this unidentified cation.

PEA medium is a heart infusion agar (Difco) with added 0.3% phenylethyl alcohol originally designed for isolation of anaerobes. The origin of its proposed use for differentiating *B. anthracis* and *B. cereus* is obscure and there is some disagreement in texts that mention it as to which of these two is selected against.

Re-examination of these media as selective agents for *B. anthracis* as against *B. cereus* is underway. Preliminary findings using a dozen culture collection strains of *B. anthracis* and one other isolate and a number of random isolates of *B. cereus* and *B. thuringiensis*, have shown that PLET medium is most promising in terms of permitting growth of *B. anthracis* and inhibiting that of *B. cereus*. Some selection in favour of *B. anthracis* is apparent in the trimethoprim-containing medium (which also contains 5 mg/litre of colistin).

These studies (to be published) have been done with dilutions of pure broth cultures; the effect of the selective agents on spore germination by the respective species and the merits of the media in the field – such as selective isolation from soils in enzootic areas – have yet to be assessed.

PEA medium has not yet been adequately tested to make any definitive assessment of its selective attributes. However, *B. cereus* and *B. anthracis* seemingly grow equally well when directly plated on to this medium.

Morphological Group 1

Defining characteristics

Sporangia (bacterial cells) not swollen, or only very slightly swollen by the spores.
Spores ellipsoidal/cylindrical, central/terminal.
Gram-positive; occasionally Gram variability reported.

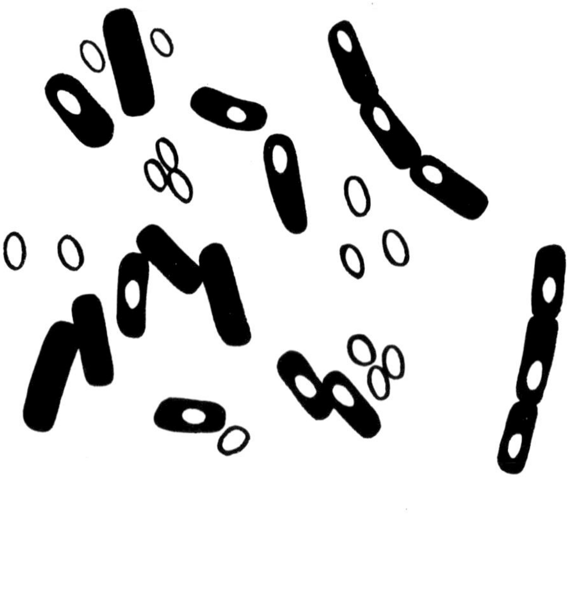

Group 1 species

1. *B. megaterium*
2. *B. cereus*
3. *B. cereus* var. *mycoides*
4. *B. anthracis*
5. *B. thuringiensis*
6. *B. licheniformis*
7. *B. subtilis*
8. *B. pumilus*
9. *B. firmus*
10. *B. coagulans*

26

A

B

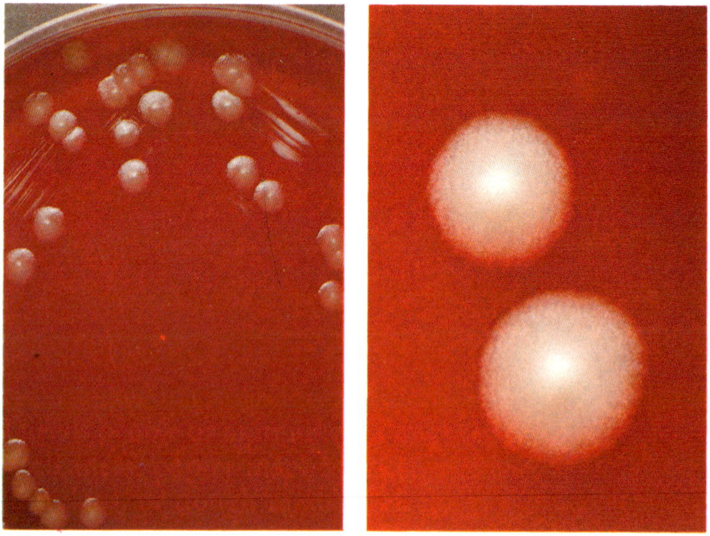

A, ×1. **B**, ×6.

Figure 26 ***B. megaterium*** **NCTC 10342/ATCC 14581 (Type strain).** 24 hours; 37°C.

Bacillus megaterium

B. megaterium is the largest-celled *Bacillus* species represented in the atlas. Its rods average 1.3 × 3.5 μm and carry oval spores, centrally or terminally placed, which do not swell the sporangium.

Sporulating cells resemble those of *B. cereus* seen in **Figure 39**. Photographs of sporulating cells of the Type strain, NCTC 10342/ ATCC 14581, whose colonies are seen in **Figure 26**, can be seen in Figure 1 of the manual of Gordon *et al* (1973).

Typical colonies measure 2–4 mm in diameter with an entire or slightly crenated edge and low convex elevation, although the Type strain (**Figure 26**) gave rise to a noticeably conical colony.

27

A B

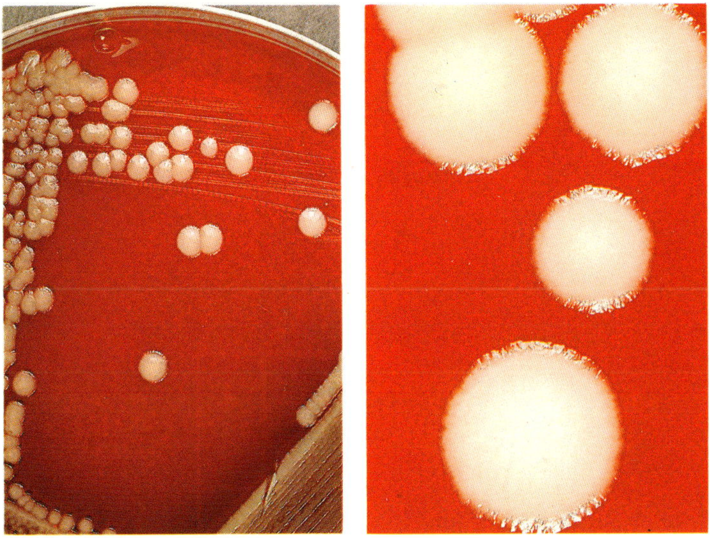

A, ×1. B, ×6.

Figure 27 ***B. megaterium*** **NCTC 6094.** 24 hours; 37°C.

The surface of the colony varies in appearance from firm and finely granular (**Figures 26**, **27** and **29**) to moist or mucoidal (generally only when the strain is one which produces large colonies).

The moist/mucoidal type of colony is seen in **Figure 28**.

Colony colour ranges from creamy-grey to grey-green.

Slight haemolysis is usually visible at 24 hours becoming marked in older cultures.

Although strictly beyond the scope of this atlas, the species (like most well-represented bacterial species, in fact) is composed of a series of strains forming a broad and continuous spectrum of properties. Gordon *et al* (1973), however, recognized two particular and broadly separable groups joined by strains with intermediate characteristics. In depth studies on DNA homology and certain phenotypic characteristics have revealed that *B. megaterium* can be subclassified into some 4 genetically differentiable subspecies (Hunger and Claus, 1981).

28

A

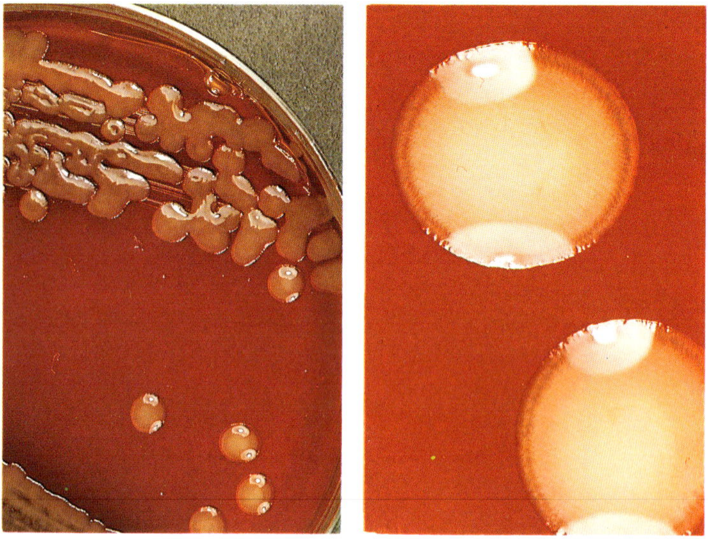

B

A, ×1. **B**, ×6.

Figure 28 ***B. megaterium*** **NCTC 5635.** 24 hours; 37°C.

The colonies of some strains can strongly resemble and be mistaken for those of *B. cereus* (**Figure 30**). 'Medusa Head' colonies with a microscopic (Gram stain) appearance similar to *B. anthracis* are sometimes encountered (Candeli *et al*, 1979). However, *B. megaterium* and the *B. cereus* group can be distinguished by the different growth and biochemical characteristics indicated in **Figure 1** and **Tables 1** and **2**. More sophisticated analysis, such as DNA base composition studies, have confirmed their separateness (Candeli *et al*, 1979).

The large size of *B. megaterium* cells and their sensitivity to lysozyme and bacteriophages have made these organisms favourites for studies of the bacterial cell, its surface, wall, cytoplasmic membranes, spores and chemical composition and also of protoplasts (Gordon *et al*, 1973).

The ATCC catalogue (1982) lists numerous strains of *B. megaterium* that have found application in commercial fermentation processes for production of a wide variety of organic chemicals together with two strains that have been put to use in assays of cycloserine and aflatoxin.

29

A

B

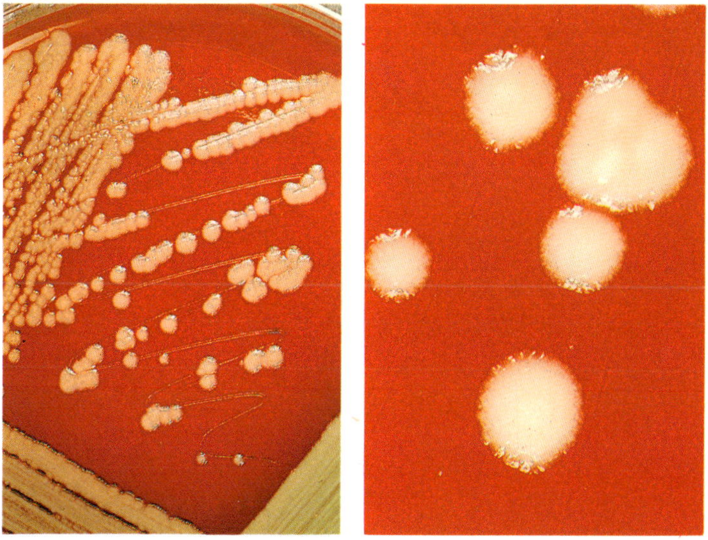

A, ×1. **B**, ×6.

Figure 29 ***B. megaterium*** **NCTC 9848.** 24 hours; 37°C.

30

A

B

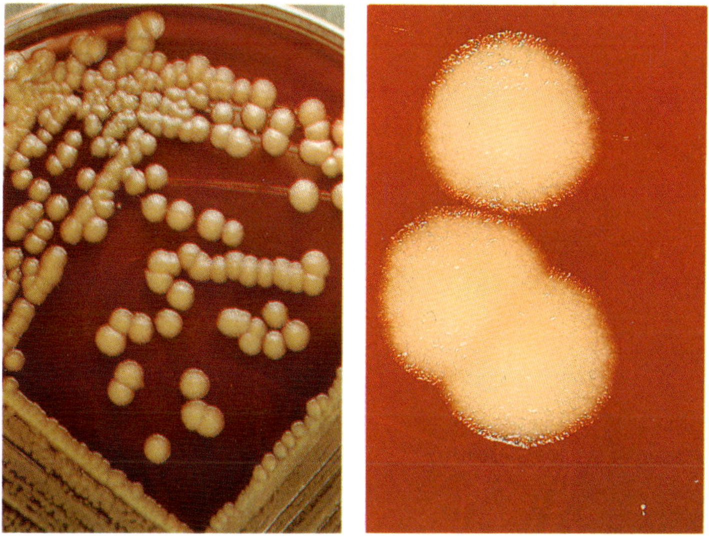

A, ×1. **B**, ×6.

Figure 30 ***B. megaterium*** **NCTC 6005.** 24 hours; 37°C.

31

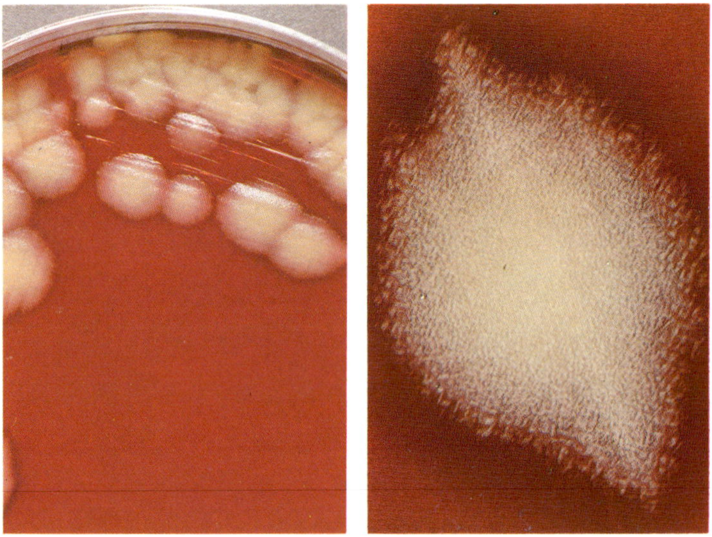

A, ×1. B, ×6.

Figure 31 ***B. cereus*** **NCTC 2599/ATCC 14579 (Type strain).** 24 hours; 37°C.

Bacillus cereus

It is our opinion (Turnbull *et al*, 1979; Turnbull, 1981) that the importance of this organism is under-recognized in the clinical laboratory. This point is underscored in Appendix 1 (*Bacillus* species as pathogens). The organism is capable of causing unpleasant infections in man and animals.

It is one of the fascinating things about *B. cereus* that, although an enormously (possibly uniquely) wide range of colonial morphologies are produced by members of the species as a whole and almost no two strains produce indistinguishable colonies, yet, even to the relatively uninitiated, the colonies when seen are usually unmistakable as probable *B. cereus*. (A reminder is necessary here that the very closely related *B. anthracis* and *B. thuringiensis* produce *B. cereus*-type colonies as discussed under **Figures 43** and **46**.

32

A

B

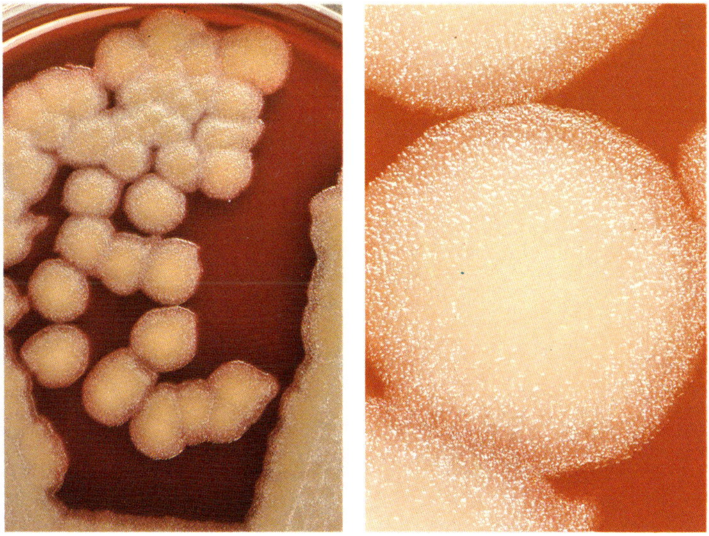

Figure 32 ***B. cereus* T.** 24 hours; 37°C.
A, ×1. Extensive haemolysis is apparent from the transparent appearance of the blood agar.
B, ×6.

Insofar as 'typical' colonies of *B. cereus* can be illustrated, **Figure 32** does so to the extent that is possible. For the species in general, after 24 hours at 37°C on conventional blood agar, colonies range from 2 to 7 mm in diameter, usually about 5 to 6 mm.

Generally, they have a medium (**Figure 32**) to coarse (**Figure 38**) matt appearance, although finely matt (**Figure 31**) or even moist smooth (**Figure 37**) surfaces are found.

B. cereus T (**Figure 32**) is a strain which produces terminal spores in the sporangia; it is sometimes referred to as *B. cereus* var. *terminalis*. This strain has been used extensively in research on a wide variety of aspects of sporulation such as heat and radiation resistance, metabolic processes and metabolites involved, the genetic control and the effects of mutagens, phages and environmental conditions. The strain was originally isolated from soil in Illinois and is not a laboratory-derived mutant.

33

A

B

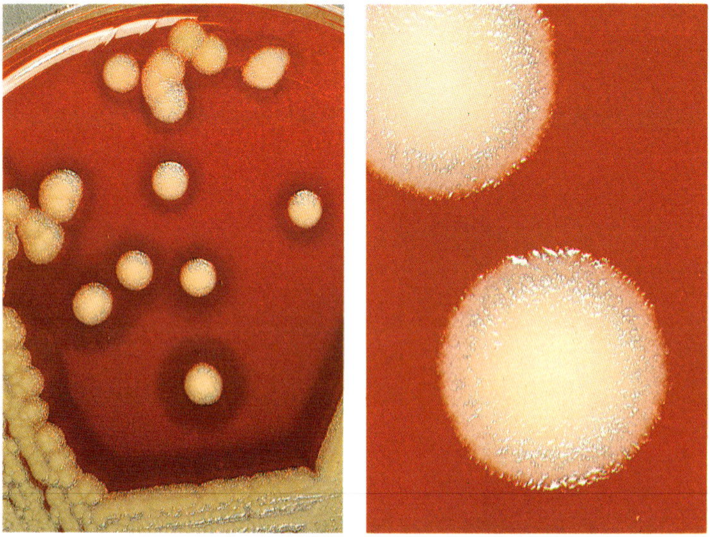

Figure 33 ***B. cereus* F4434/73.** 24 hours; 37°C.
A, × 1. Note the zones of haemolysis around the colonies.
B, × 6.

The colonies in **Figure 33** are also fairly 'typical' *B. cereus* colonies although somewhat smaller than average.

In general, colonies range from circular with an entire edge (**Figure 33**) to irregular and fimbriate (**Figures 36** and **37**). All have a low-convex to convex elevation and many are markedly haemolytic. Colonies on blood agar frequently have a greenish tinge probably acquired from a haematogenous pigment derivative of haemoglobin produced during some haemolytic reactions.

Cultures have been described by laboratory workers as having the smell of 'crushed strawberries' or 'musty books'.

An unusual phenomenon which we have noted to occur with many, but apparently not all strains of *B. cereus* is the development upon ageing of marked colonial variation (**Figure 35**). The colonial variants, if plated out and incubated at 37°C for 24 hours, become indistinguishable again. Where the strain is serotypable, the variants are all of the same serotype. We have not studied the phenomenon in depth and have not seen other references to it. The possibility that it may be a form of 'Dienes' phenomenon' (Dienes, 1946) is probably worth investigating.

34

A

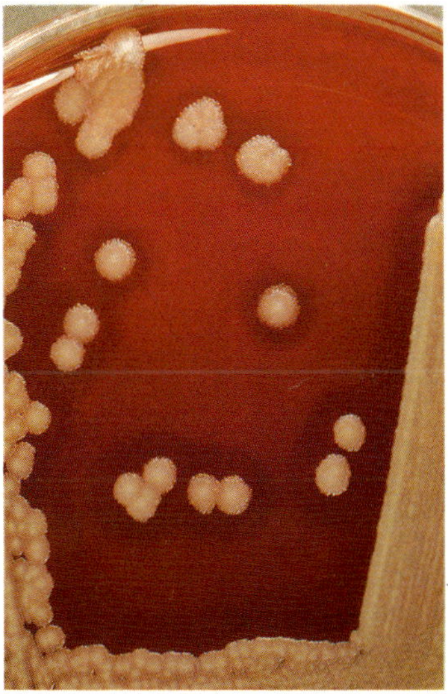

B

A, ×1. **B**, ×6.

Figure 34 ***B. cereus* F2038/78.** 24 hours; 37°C.

35

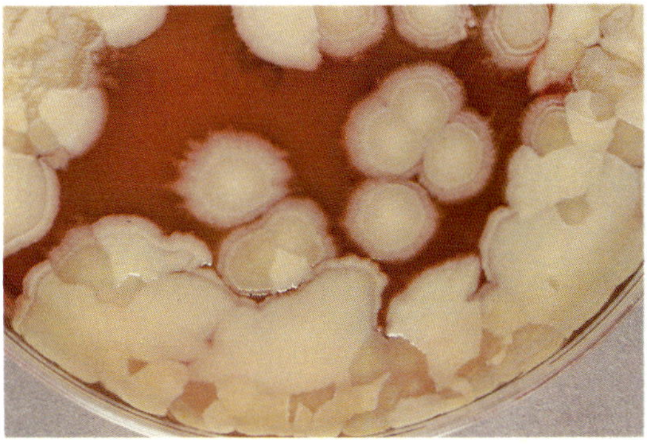

Figure 35 ***B. cereus* NCTC 2599/ATCC 14579 (Type strain).** 24 hours at 37°C followed by 1 week on the bench at room temperature. ×1.

36

A

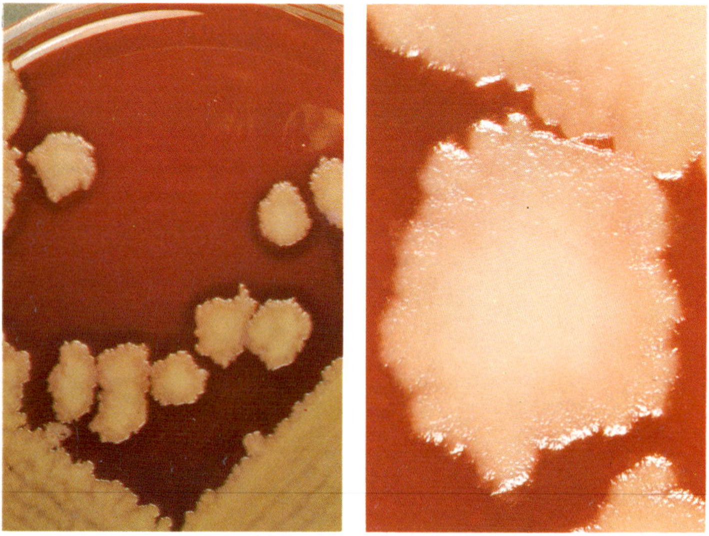

B

A, ×1. **B**, ×6.

Figure 36 ***B. cereus*** **F658/78.** 24 hours; 37°C.

37

A

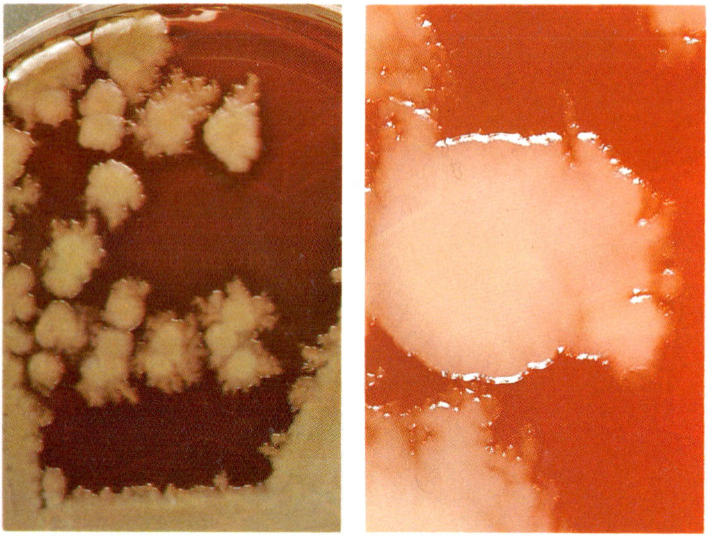

B

A, ×1. **B**, ×6.

Figure 37 ***B. cereus*** **F589/78.** 24 hours; 37°C.

38

A

B

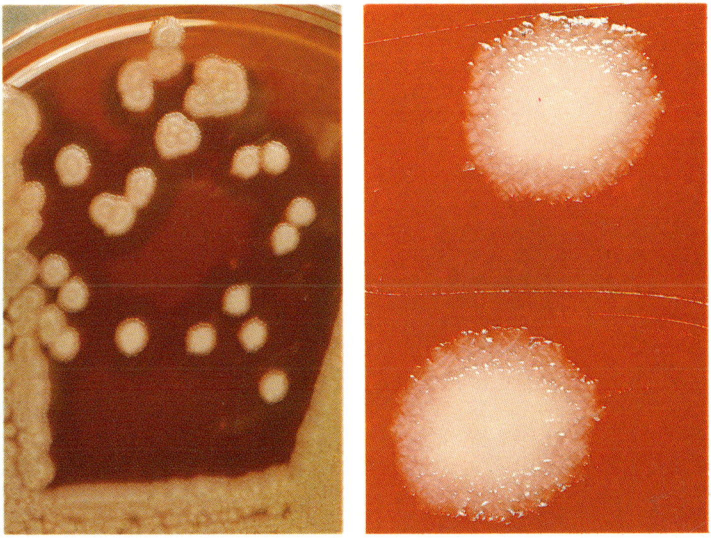

A, ×1. **B**, ×6.

Figure 38 ***B. cereus*** **F829/78.** 24 hours; 37°C.
Note the 'fried egg' appearance of these colonies.

39

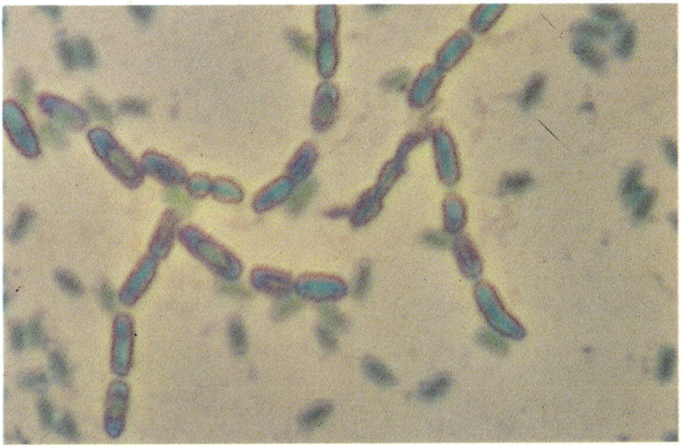

Figure 39 ***B. cereus*** **spores.** Phase contrast (Zeiss). ×1800.

B. cereus bacterial cell sizes average about $1 \times 3.5\,\mu m$. The spores are large and oval and may or may not slightly swell the sporangium. **Figure 39** was photographed through a × 40 objective and × 5 camera 'eyepiece' on to 35 mm film magnified × 9 in this final print.

***The relatedness of* B. cereus, B. anthracis, B. mycoides *and* B. thuringiensis**

Gordon *et al* (1973) noted that, while severe criticisms were levelled at their conception of *B. cereus* as a 'parent species' or 'stable species' and of *B. anthracis* and *B. thuringiensis* as varieties, the designation of *B. mycoides* as a variety of *B. cereus* was calmly received.

This, perhaps, was largely a reflection of the comparative lack of importance of, and therefore interest in, *B. mycoides*. *B. anthracis* and *B. thuringiensis* have long been organisms of particular interest – *B. anthracis* as a serious pathogen of man and animals, and *B. thuringiensis* as a commercially important insect pathogen.

In fact, Kaneko *et al* (1978) proposed that, in accordance with the Bacteriological Code, since *B. anthracis* was the earliest name applied to the species, nomenclature should be *B. anthracis* subspecies *anthracis*, *B. anthracis* subspecies *cereus*, *B. anthracis* subspecies *thuringiensis* with the same to be implied for *B. mycoides*. However, the argument of Gordon and her colleagues centred around the fact that a strain of *B. anthracis* which had lost its virulence, or a strain of *B. thuringiensis* which had lost its ability to produce parasporal crystals, if encountered in a laboratory, would be identified as *B. cereus*. Thus they regarded *B. cereus* as the 'stable' species. Further problems would arise from the proposed nomenclature of Kaneko *et al* (1978) firstly in the 'alarm' aspect of calling organisms *B. anthracis* and secondly in the cumbersome names that would result in giving the serological subtypes of *B. thuringiensis* their full titles.

It would seem from the Approved Lists of Bacterial Names (1980) that, even if the problem of correct identity has not been resolved by laboratory investigations, the International Committee on Systematic Bacteriology has now accepted that these organisms can be referred to in simple species terms – i.e. *B. anthracis* rather than *B. cereus* var. *anthracis*, *B. mycoides* rather than *B. cereus* var. *mycoides*, and the same for *B. thuringiensis*.

Nevertheless, it is important not to lose sight of the closeness of the relationship between these four species, borne out by DNA homologies (Kaneko *et al*, 1978; Priest, 1981) in addition to their essentially indistinguishable biochemical reactions (**Table 2**).

Bacillus mycoides (B. cereus var. *mycoides)*

This rhizoid relation of *B. cereus* produces large colonies, 3–8 mm in diameter, which frequently cover the surface of the plate with their outgrowths (**Figure 40**).

The colonies are flat, sometimes extremely crenated with the characteristic outgrowths from their edges (**Figures 40** and **41**). They frequently have a grey-green tinge similar to that of *B. cereus* on blood agar.

Older cultures show marked haemolysis and are sometimes difficult to subculture as they adhere to the agar.

A principal difference from *B. cereus* is that many *B. mycoides* strains are non-motile.

40

A  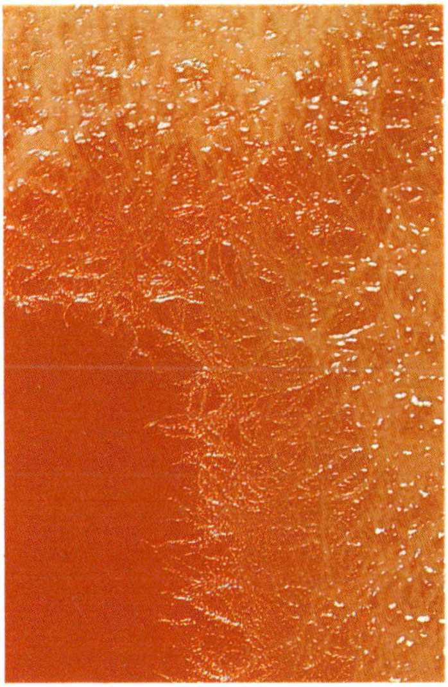 B

A, ×1. B, ×6.

Figure 40 ***B. mycoides (B. cereus*** **var.** ***mycoides)*** **F4894/78.**
24 hours; 37°C

41

A 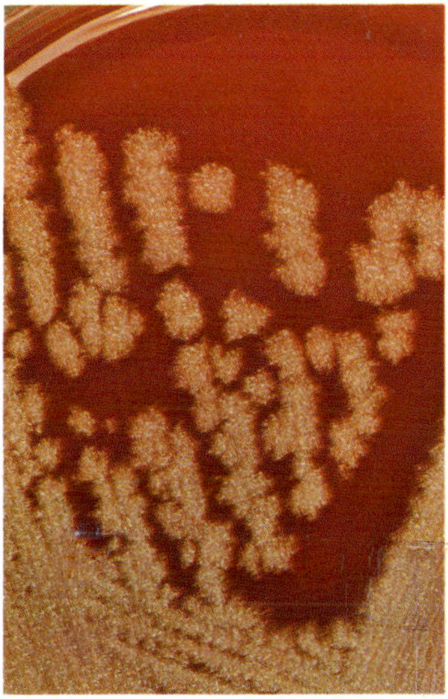 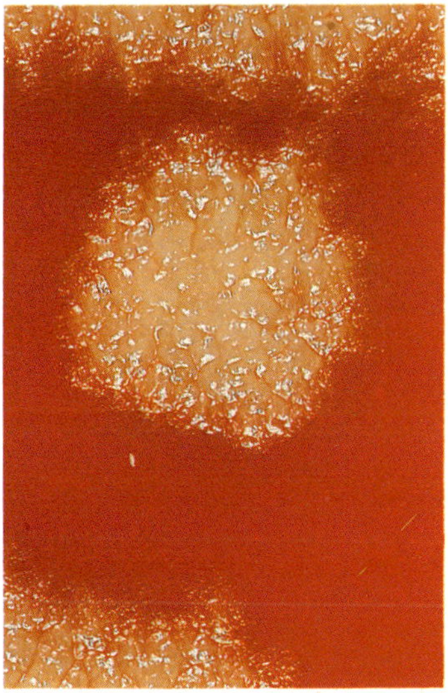 B

A, ×1. B, ×6.

Figure 41 ***B. mycoides (B. cereus*** **var.** ***mycoides)*** **F4598/76.**
24 hours; 37°C.

42

A

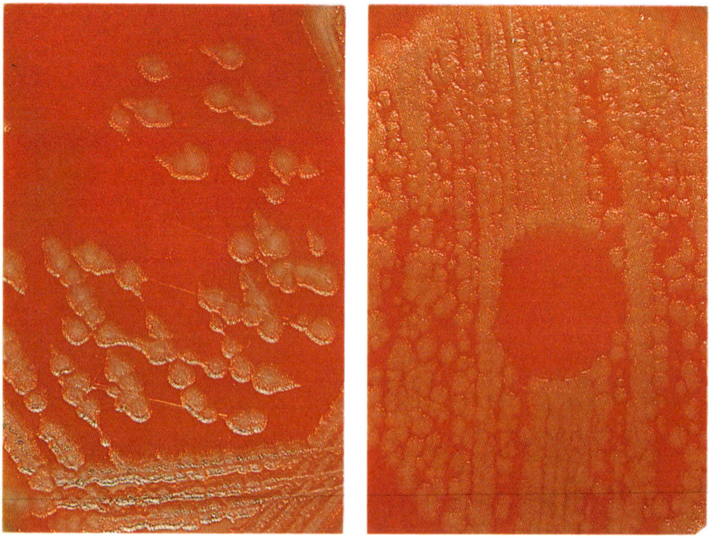

B

A, ×1. **B**, Lysis by gamma-phage. ×1.

Figure 42 ***B. anthracis.*** 24 hours; 37°C.

Bacillus anthracis

Extensive coverage has already been given (Methods and Characterization Tests, section VIII) to ways by which *B. anthracis* may be distinguished from *B. cereus* and **Figures 21–24** have already highlighted a number of the principal features of *B. anthracis*.

Figure 42A (kindly supplied by the Public Health Laboratory, Guildford, at that time, the UK anthrax reference laboratory*), like **Figure 21**, shows the spiked colonies of another strain. **Figure 42B** shows the susceptibility of this strain to γ-phage.

*Anthrax reference facilities in the United Kingdom have now been transferred to the Vaccine Research and Production Laboratory, PHLS–CAMR, Porton Down, Salisbury, Wiltshire SP4 0JG.

43

A

B

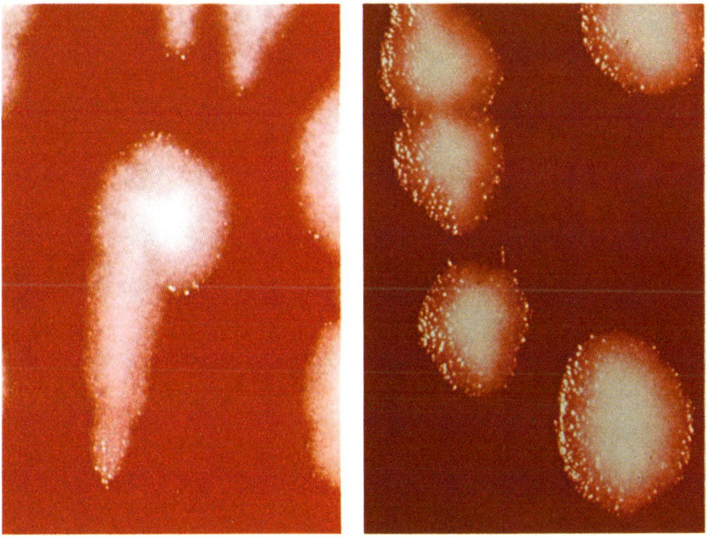

A, × 6. Spiked colonies. **B**, × 6. Non-spiked colonies.

Figure 43 ***B. anthracis*** **A19432/80.** 24 hours; 37°C.

Figure 43 shows further colonies of strain A19432/80 (see **Figure 21**). The similarity between *B. anthracis* and *B. cereus* colonies is clear in the non-spiked colonies (**Figure 43B**).

As described previously, lack of haemolysis, whiteness of the colonies and marked tackiness on probing with a loop or needle are initial warning signs that the organism may be *B. anthracis*. Further tests should then be carried out as described under Methods and Characterization Tests, section VIII.

44

A 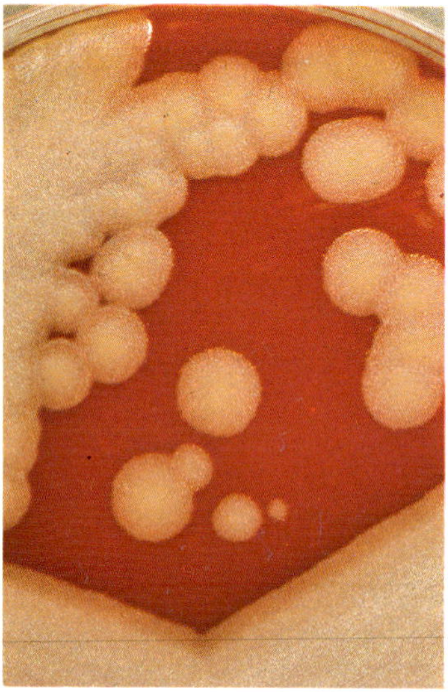 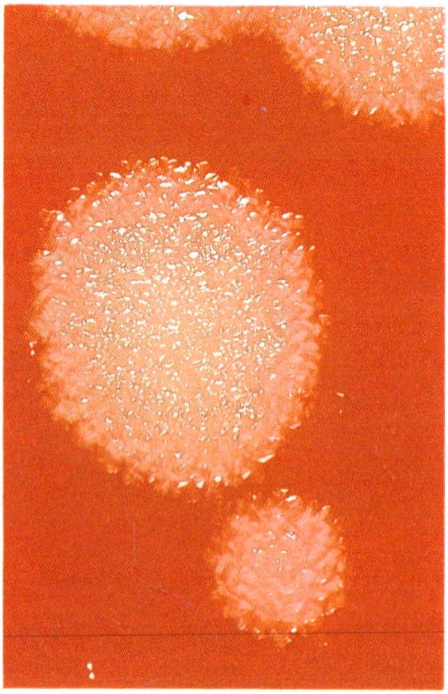 B

A, ×1. **B**, ×6.

Figure 44 ***B. thuringiensis* – example 1.** 24 hours; 37°C.

Bacillus thuringiensis

The close relationship between *B. thuringiensis* and *B. cereus* was discussed on page 104 and the importance of *B. thuringiensis* in insect control was pointed out in the section on parasporal crystal bodies (Methods and Characterization Tests, section III.3.c). As a result of its insecticidal importance, the organism has been studied as extensively as *B. cereus* and *B. anthracis*.

Principally on the basis of flagellar antigen typing (Methods and Characterization Tests, section VII), *B. thuringiensis* has been subdivided by the insect pathologists into some 19 subspecies. However, serotyping can be done only by centres holding the specific antisera; in the UK, the reference centre for *B. thuringiensis* is the Glasshouse Crops Research Institute, Insect Pathology Group, Worthing Road, Rustington, Littlehampton, West Sussex BN16 3PU.

45

A

B

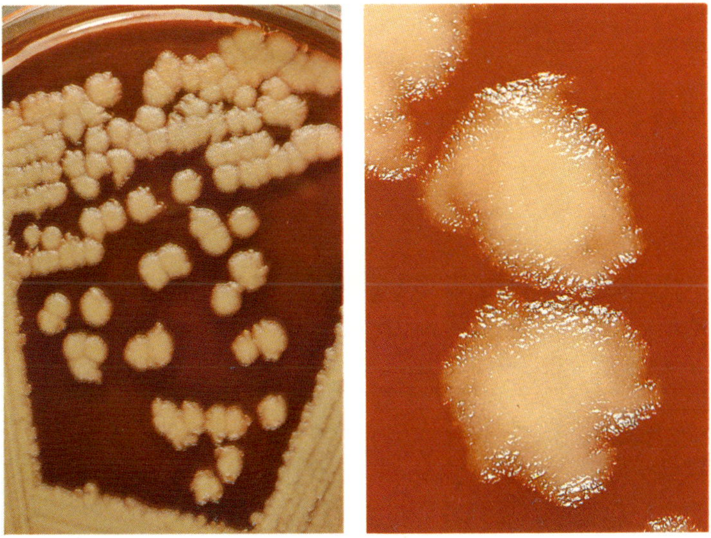

A, ×1. **B**, ×6.

Figure 45 ***B. thuringiensis* – example 2.** 24 hours; 37°C.

Few laboratories that attempt to identify *Bacillus* species, on encountering a *B. cereus*-like isolate, consider looking for the parasporal crystal which is the principal differential characteristic for identifying the isolate as *B. thuringiensis* rather than *B. cereus*. They are certainly to be encouraged to do so.

Colonially (**Figures 44–46**) and biochemically (**Table 2**), the two species are indistinguishable.

46

A

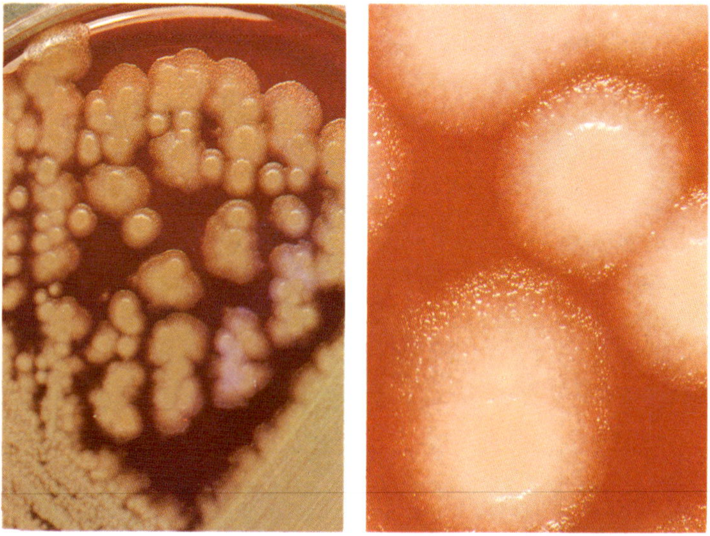

B

A, ×1. B, ×6.

Figure 46 ***B. thuringiensis*** **– example 3.** 24 hours; 37°C.

Since *B. thuringiensis* is used in commercial insecticidal crop sprays in many parts of the world, it might be expected that the bacterium would be encountered in laboratories of such regions or elsewhere from residual spores on crops that have been harvested and transported.

In fact, only a single report has come to our notice of a clinical isolation of *B. thuringiensis* and this was related to the accidental inoculation of the web space of a finger in a laboratory worker with a concentrated spore and toxin preparation. The preparation was also contaminated with *Acinetobacter calcoaceticus*. Severe inflammation and lymphangitis did result from this infection (Public Health Laboratory Service, Communicable Disease Report, No. 1, 1982).

Of a representative number of isolates from food poisoning and other clinical cases identified in the Food Hygiene Laboratory as *B. cereus*, none were found to be *B. thuringiensis* (Dr H.D. Burges, personal communication).

The *Bacillus subtilis* group

B. licheniformis, *B. subtilis* and *B. pumilus* comprise what taxonomic experts refer to as the 'subtilis group' or 'subtilis spectrum' (Gordon *et al*, 1973; Gordon, 1975; Logan and Berkeley, 1981).

As implied by these terms, the three species are closely related physiologically, biochemically and as shown by DNA reassociation studies and pyrolysis gas liquid chromatography (Logan and Berkeley, 1981).

As seen in **Table 2**, there are a few tests by which they may be distinguished but Gordon (1975) surmises that the day may come when it is considered purposeless to do so and that they are lumped together.

The cells of *B. licheniformis*, *B. subtilis* and *B. pumilus* are generally of smaller dimensions than those of *B. megaterium* and the *B. cereus* group.

As reviewed in Appendix 1, *B. licheniformis* and *B. subtilis* have been implicated on occasion as food poisoning agents and all three species have been associated with other types of infections, at least opportunistically.

They are also occasional agents of spoilage in canned foods, milk and dairy products and the cause of 'ropiness' in bread (Norris *et al*, 1981).

B. licheniformis and *B. subtilis* are of considerable importance in a number of microbiological, chemical and industrial fields such as the assay of certain antibiotics, production of certain organic compounds including the antibiotic bacitracin and the enzyme penicillinase, and sterilization tests. This is covered a little further in Appendix 3. *B. subtilis* has also been the organism of choice for considerable amounts of research to promote understanding of metabolic and genetic processes.

47

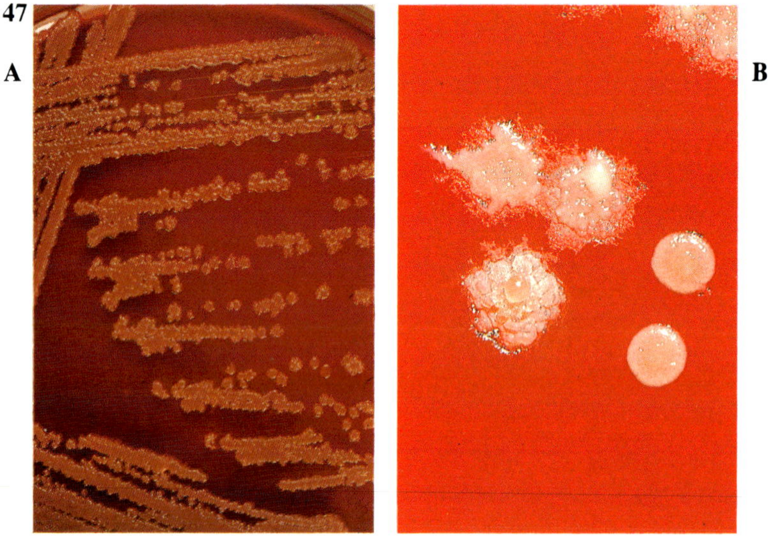

A, ×1. B, ×6.

Figure 47 ***B. licheniformis*** **NCTC 10341/ATCC 14580 (Type strain).** 24 hours; 37°C.

Bacillus licheniformis

Colonies vary from 2 to 4mm in diameter with an outline that may be undulated or crenated and an elevation that tends to be convex with a papillate surface.

Small amounts of mucous fluid may be seen in association with the colonies of some strains when the cultures are young (**Figure 47B**). This evaporates readily upon exposure to the atmosphere.

Typical colonies adhere closely to the agar and the species name results from this property. The colonies are haemolytic with colour varying from creamy-buff to grey-green.

The principal distinguishing feature between *B. licheniformis* and *B. subtilis* which can produce very similar colonies (**Figures 55–58**) is that *B. licheniformis* grows readily anaerobically whereas *B. subtilis* does not.

Utilization of propionate (*B. licheniformis* +; *B. subtilis* −; see **Figure 19**) is another useful distinguishing test. Also, *B. licheniformis* liquefies gelatin slowly (4 days – 3 weeks).while *B. subtilis* usually does so within 2–3 days.

48

A

B

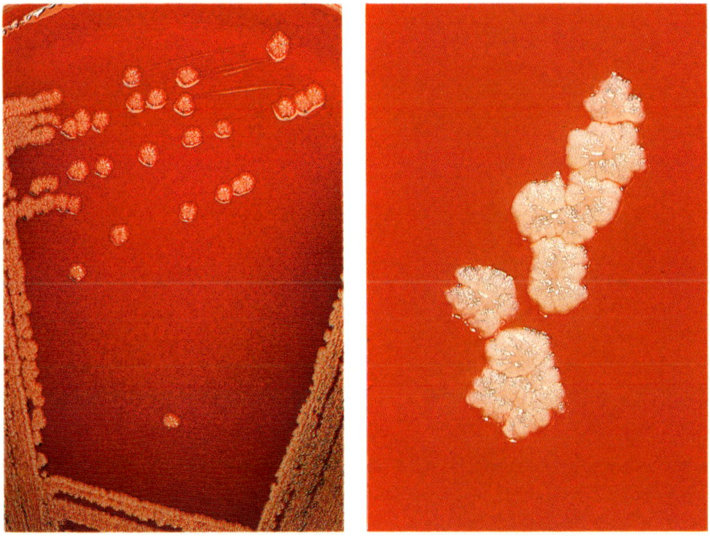

A, ×1. **B**, ×6.

Figure 48 ***B. licheniformis*** NCTC **9932.** 24 hours; 37°C.

49

Figure 49 ***B. licheniformis*** **NCTC 2120** (formerly also ATCC 8187). ×6. 24 hours; 37°C.

50

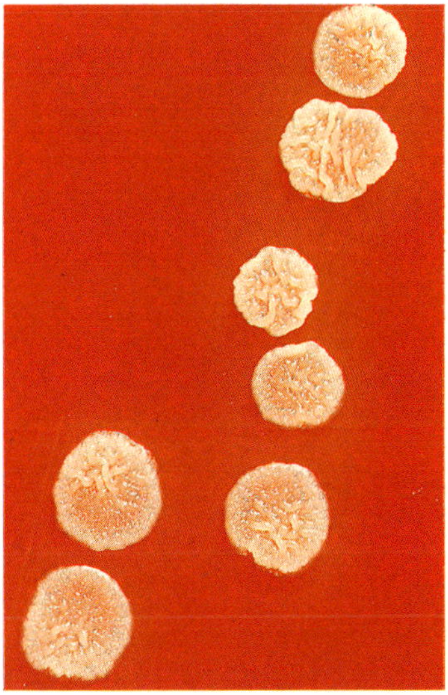

Figure 50 ***B. licheniformis*** **NCTC 8233/ATCC 13438.** ×6. 24 hours; 37°C.

51

A

B

A, ×1. **B**, ×6.

Figure 51 ***B. licheniformis*** **NCTC 8720.** 24 hours; 37°C.

52

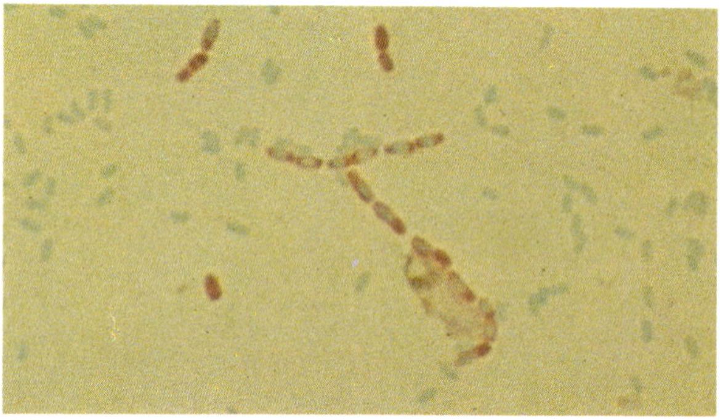

Figure 52 **Spores of *B. licheniformis*** (malachite green stain; ×1200).

Figures 53 and **54** show less typical colonies.

The bacterial cell size averages $0.7 \times 2.2\,\mu m$. The spores are oval, central or subterminal (**Figure 52**) and only swell the sporangium slightly if at all.

53

A B

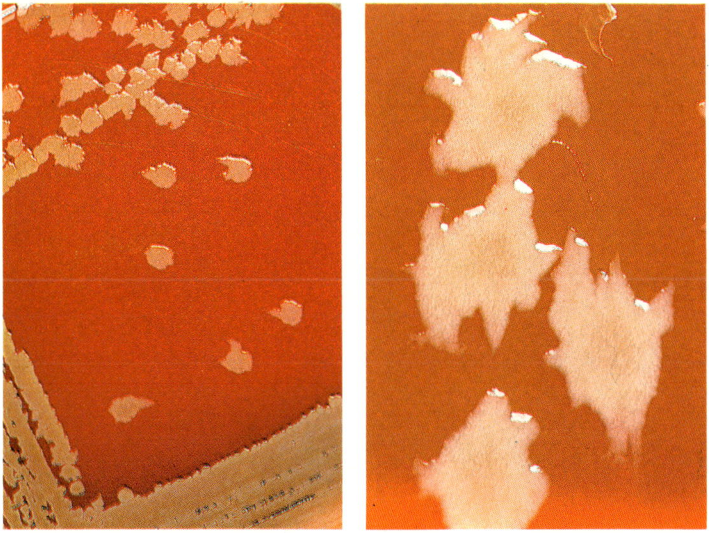

A, ×1. **B**, ×6.

Figure 53 ***B. licheniformis*** **NCTC 7589.** 24 hours; 37°C.

54

A B

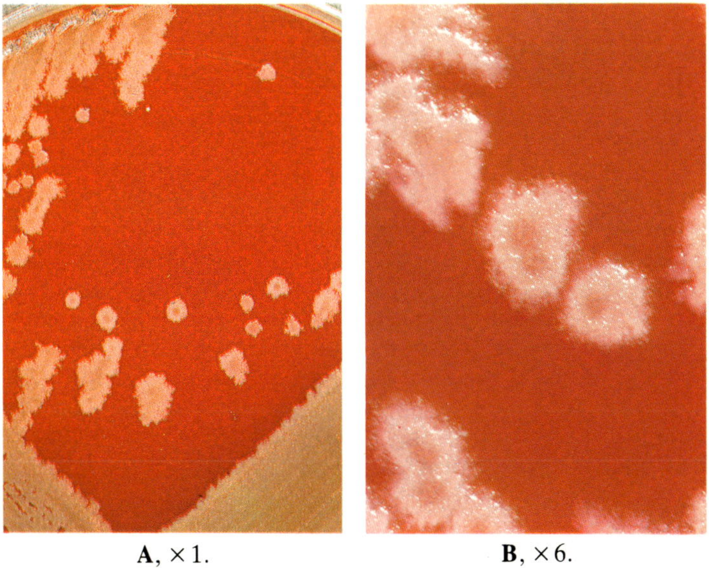

A, ×1. **B**, ×6.

Figure 54 ***B. licheniformis*** **NCTC 6346.** 24 hours; 37°C.

55

A

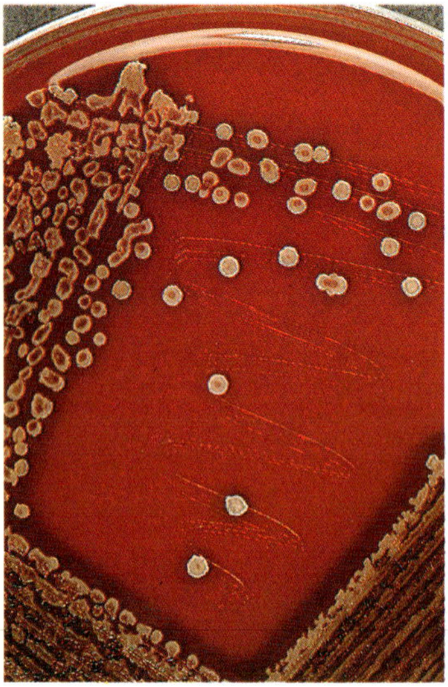

B

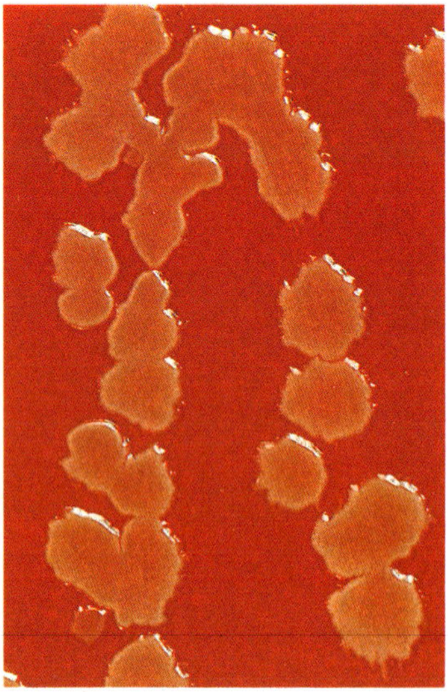

A, ×1. **B**, ×6.

Figure 55 ***B. subtilis*** **NCTC 3610/ATCC 6051 (Type strain).** 24 hours; 37°C.

Bacillus subtilis

In terms of nomenclature, this is the oldest species of the genus *Bacillus*.

A range of colonial morphologies is found within the species, from moist butyric (**Figures 57** and **60A, B**) and mucoid (**Figure 55**) types to rough friable types (**Figures 58**, **59**, **60C** and **61**).

The moist and mucoid types tend to dry out rapidly and assume a friable or crenated appearance (compare **Figures 55** and **56**).

The two plates shown in **Figure 56** were grown and photographed on two separate occasions. The slight differences in appearance probably reflect different ages, moisture content etc. of the blood agar used on the two occasions. The strain was haemolytic; lighting conditions under which the lower plate was photographed have made this a little harder to see than in the top plate.

56

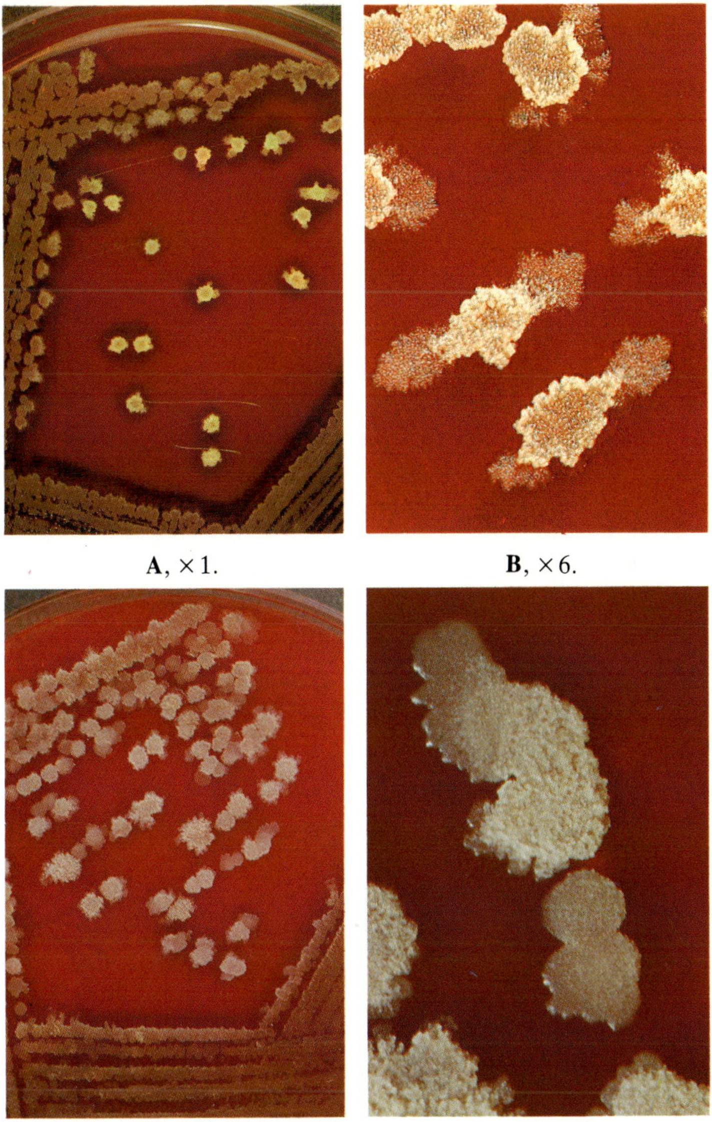

A, ×1. B, ×6.

C, ×1. D, ×6.

Figure 56 ***B. subtilis*** **NCTC 3610/ATCC 6051.** Crenated forms. 24 hours; 37°C.

57

A

B

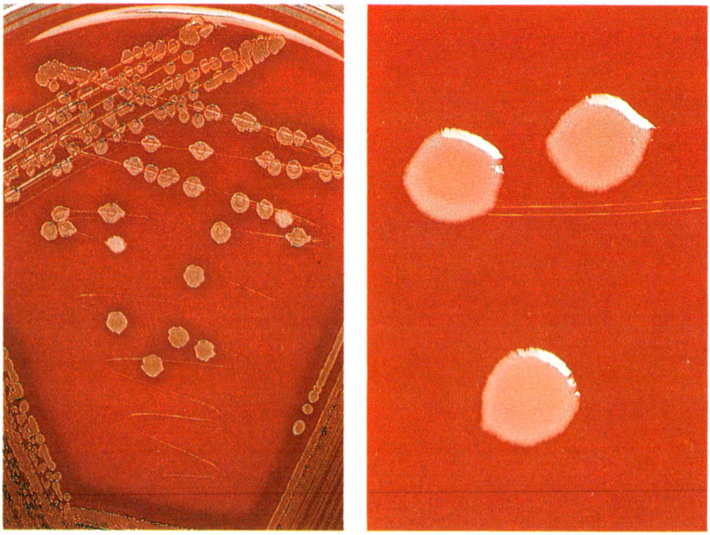

A, ×1. **B**, ×6.

Figure 57 ***B. subtilis*** **NCTC 10073 (formerly *B. globigii*).** 24 hours; 37°C.

Two colonial types were found in cultures of *B. subtilis* NCTC 10073 – a moist butyrous type (**Figure 57**) and a rough friable type (**Figure 58**). These bred true and both were confirmed as *B. subtilis*.

The two plates shown in **Figure 58** were cultured and photographed on two separate occasions. As with **Figure 56**, the differences apparent can be attributed to differences in the batch of blood agar used and, to some extent, to the lighting during photography.

In general, colonies of *B. subtilis* have a convex elevation and edges which may be entire, crenated or undulated. They are generally not adherent to the agar surface and can be picked off easily for subculture. Haemolysis is produced to varying degrees.

58

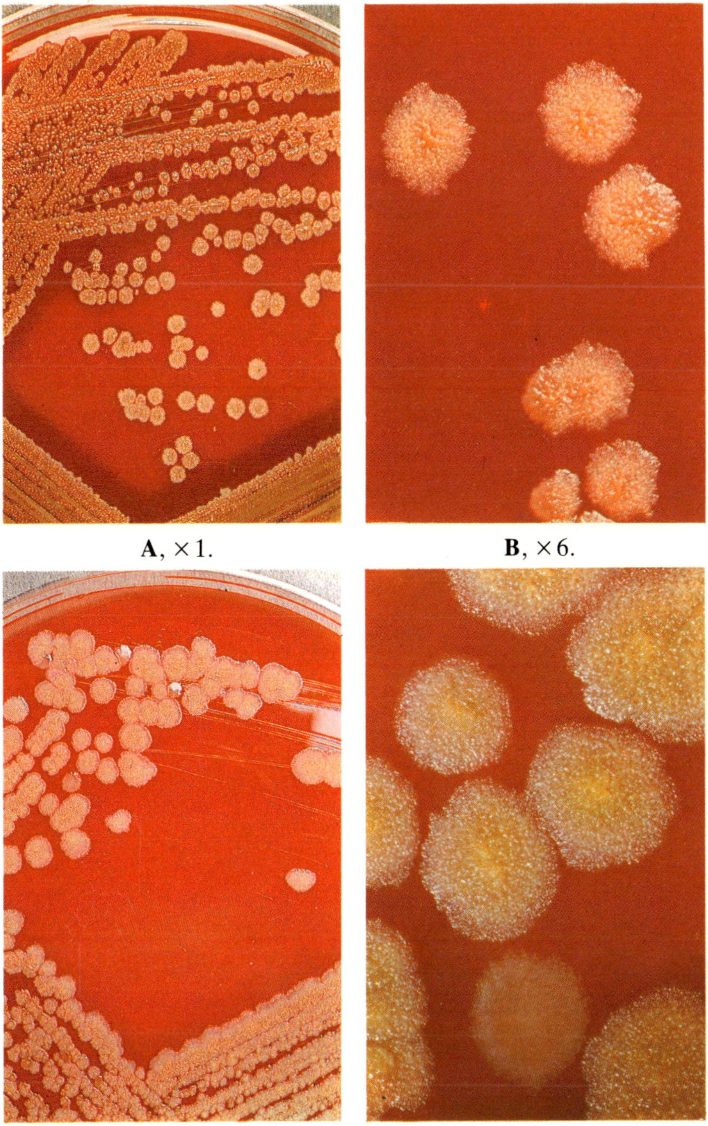

A B

A, ×1. B, ×6.

C D

C, ×1. D, ×6.

Figure 58 ***B. subtilis*** **NCTC 10073 (rough type).** 24 hours; 37°C.

59

A

B

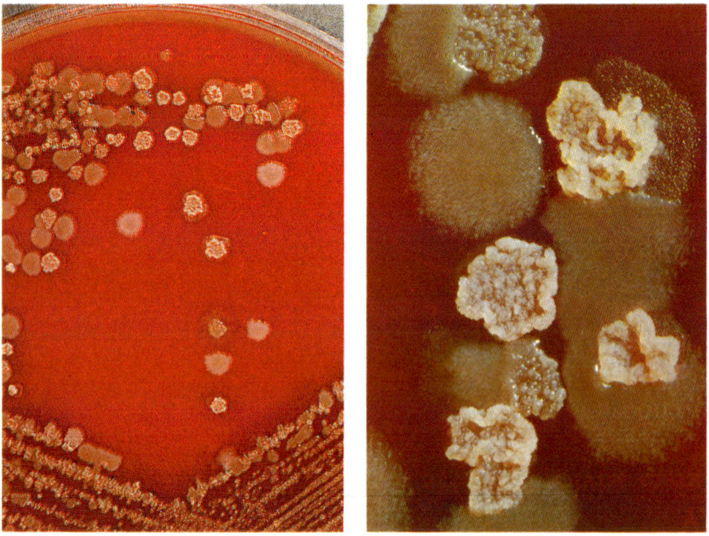

A, ×1. **B**, ×6.

Figure 59 ***B. subtilis*** **NCTC 5398.** 24 hours; 37°C.

Moist butyric or mucoid type colonies in various stages of drying often give the bizarre sort of 'mixed' appearance seen in **Figures 59**, **62** and **63**.

In fact, two colony types which bred true (both confirmed as *B. subtilis*) were found in cultures of strain NCTC 5398 (**Figure 60**) – a moist butyrous type (top) and a friable rough type (bottom). Strains 6432 (**Figure 62**) and 7861 (**Figure 63**) were apparently pure cultures.

Strains NCTC 3610 (**Figures 55** and **56**), NCTC 10073 (**Figures 57** and **58**) and NCTC 10040 (**Figure 61**) have been or are being put to use in assays of metabolites and antibiotics – for instance, in milk and pharmaceutical products – and in chemical (e.g. ethylene oxide) sterilization tests.

B. subtilis NCTC 6432 (**Figure 62**) featured in the early work on antibiotics in the early 1940s.

60

A

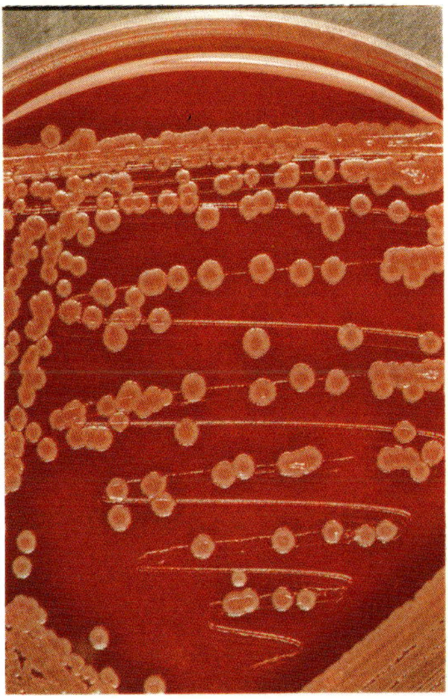

B

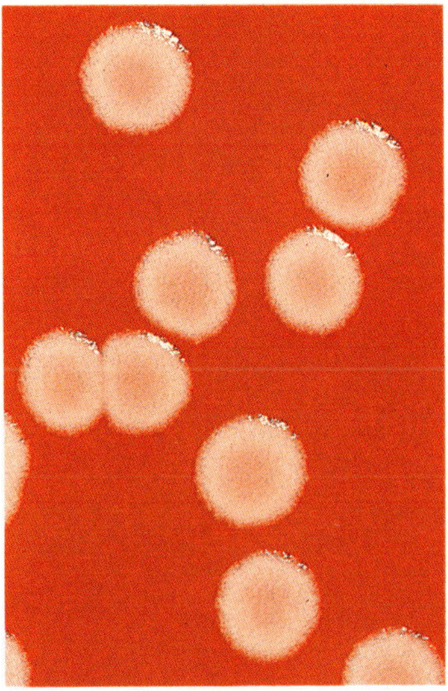

A, ×1.

B, ×6.

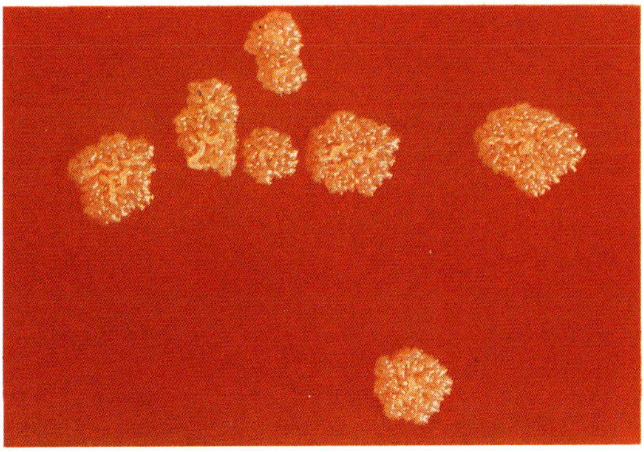

C

C, ×6.

Figure 60 ***B. subtilis*** **NCTC 5398. Two colony types.**
24 hours; 37°C.

61

A

B

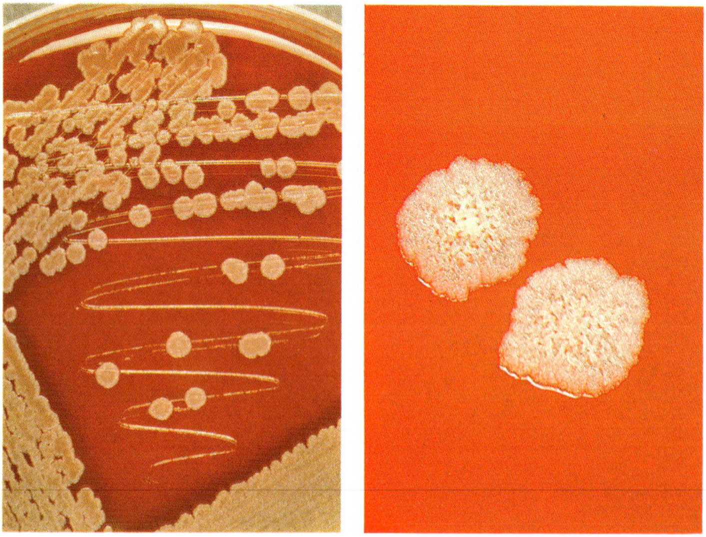

A, ×1. **B**, ×6.

Figure 61 ***B. subtilis*** **NCTC 10400/ATCC 6633.** 24 hours; 37°C.

62

A

B

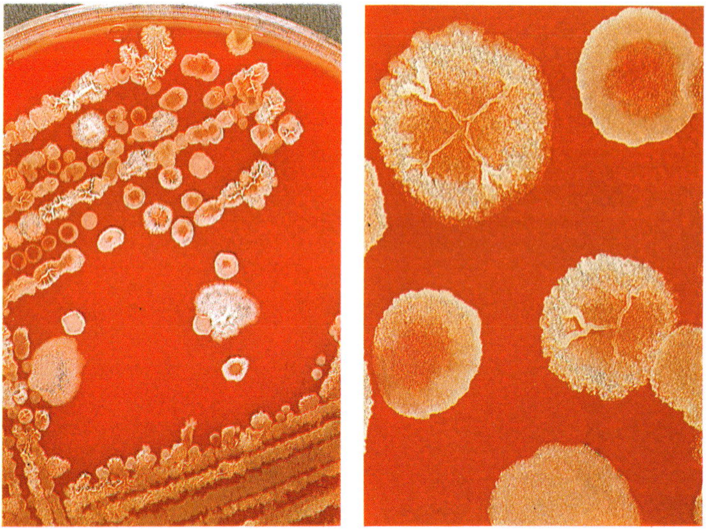

A, ×1. **B**, ×6.

Figure 62 ***B. subtilis*** **NCTC 6432.** 24 hours; 37°C.

63

A

B

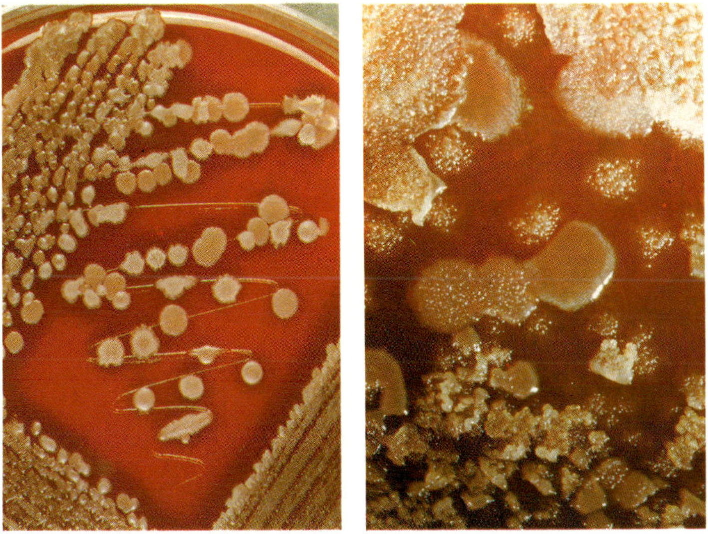

A, ×1. **B**, ×6.

Figure 63 ***B. subtilis*** **NCTC 7861 (formerly *B. subtilis* var. *niger*).** 24 hours; 37°C.

64

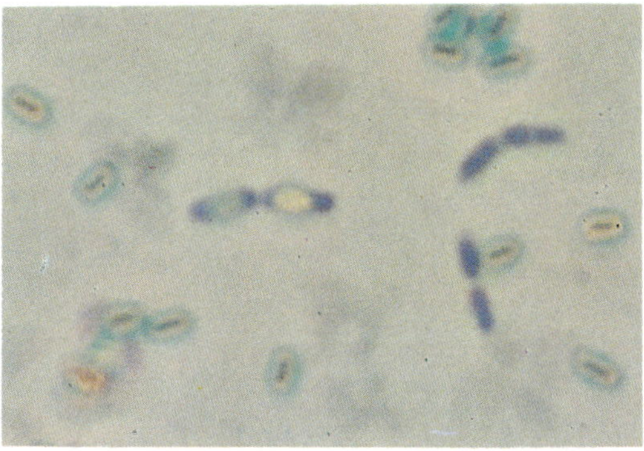

Figure 64 ***B. subtilis*** **spores.** Phase contrast.

B. subtilis cells average $0.7 \times 2.5\,\mu m$ and contain oval, central or subterminal spores which may just slightly distend the sporangium (**Figure 64**).

65

A

B

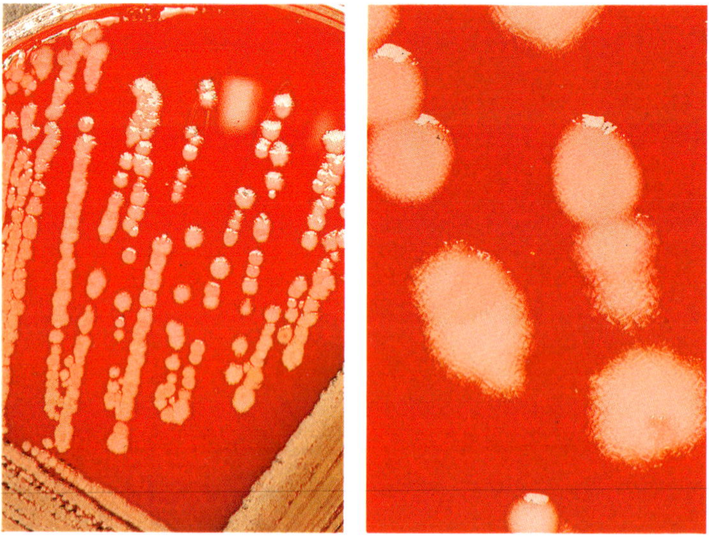

A, ×1. **B**, ×6.

Figure 65 ***B. pumilus*** **NCTC 10337/ATCC 7061 (Type strain).**
24 hours;37°C.

Bacillus pumilus

This is the third member of the '*B. subtilis* group' or '*B. subtilis* spectrum' (see page 113). The species is biochemically similar to *B. subtilis* except that it does not reduce nitrate to nitrite and it is unable to hydrolyse starch.

The strains we examined produced colonies which lacked the ruggedly crenated appearance we considered typical of *B. licheniformis* although they sometimes appeared slightly crenated. Generally, colonies were 3–4 mm in diameter, moist and butyrous with raised elevation and varying in colour from creamy-buff to grey.

Haemolysis was often apparent and sometimes strong (**Figure 68A**).

The cell size averages $0.6 \times 2.5\,\mu m$ and, very similar in appearance to *B. subtilis* (**Figure 64**) and *B. licheniformis* (**Figure 52**), they contain oval, central or subterminal spores which may slightly distend the sporangium.

66

A

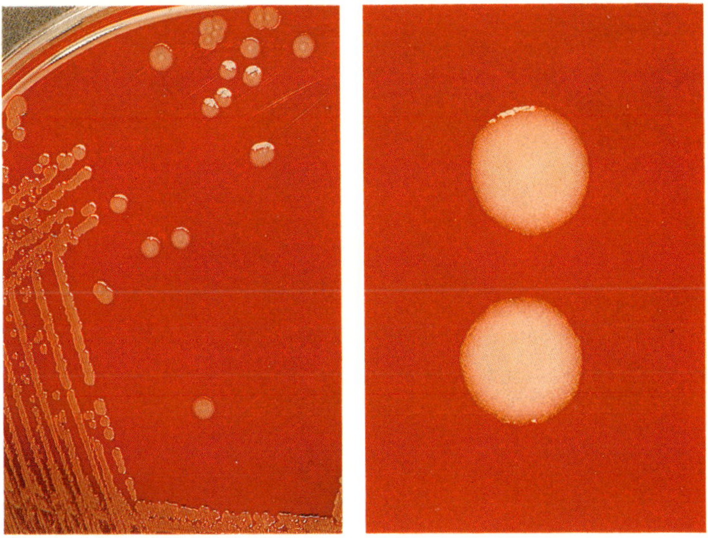

B

A, ×1. **B**, ×6.

Figure 66 ***B. pumilus*** **NCTC 2595/ATCC 4520.** 24 hours; 37°C.

67

A

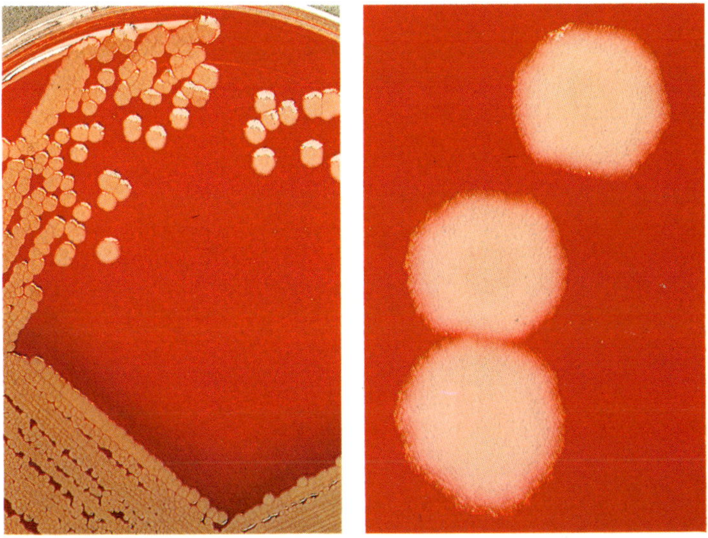

B

A, ×1. **B**, ×6.

Figure 67 ***B. pumilus*** **NCTC 8241/ATCC 14884.** 24 hours; 37°C.

68

A

B

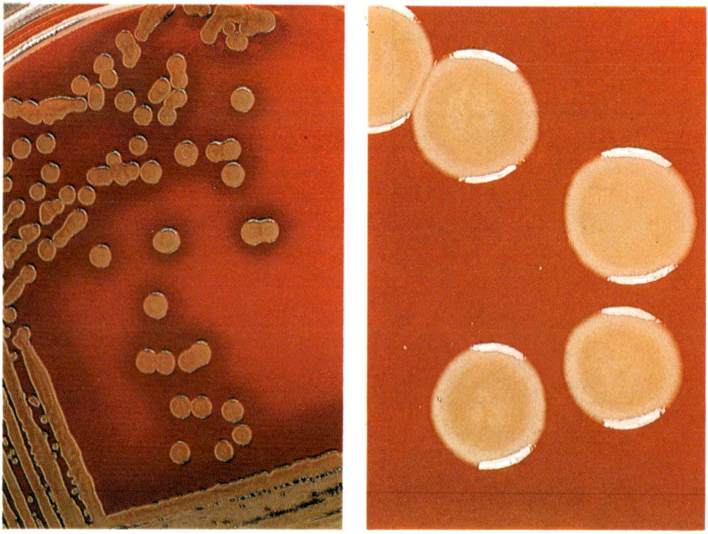

A, ×1. **B**, ×6.

Figure 68 ***B. pumilus*** **NCTC 9436.** 24 hours; 37°C.

To our knowledge, *B. pumilus* has only been definitely associated with infections on two occasions – a case of meningitis with bacteraemia (Weinstein and Colburn, 1950) and a rectal abscess (Melles *et al*, 1969) but it probably joins the ranks of many *Bacillus* species discarded daily in clinical laboratories as 'inconsequential ASBs' which, in fact, merit closer attention.

Strains of *B. pumilus* are used, like *B. subtilis*, for production of a number of organic compounds. Strain NCTC 8241/ATCC 14884 (**Figure 67**) is listed as an appropriate organism for assaying neomycin (Kavanagh, 1972) and trimethoprim (Garrod *et al*, 1981). *B. pumilus* has also found use in studies on radiation resistance of spores (Borick and Fogarty, 1967) and strain NCTC 10327 (**Figure 70**) has been used as a standard organism for testing the efficacy of sterilization procedures using radiation (Darmady *et al*, 1961).

69

A

B

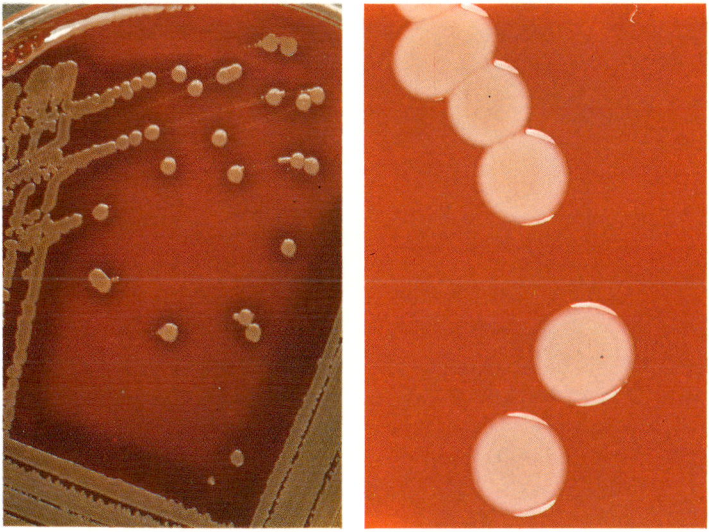

A, ×1.

B, ×6.

Figure 69 ***B. pumilus*** **NCTC 7576.** 24 hours; 37°C.

70

A

B

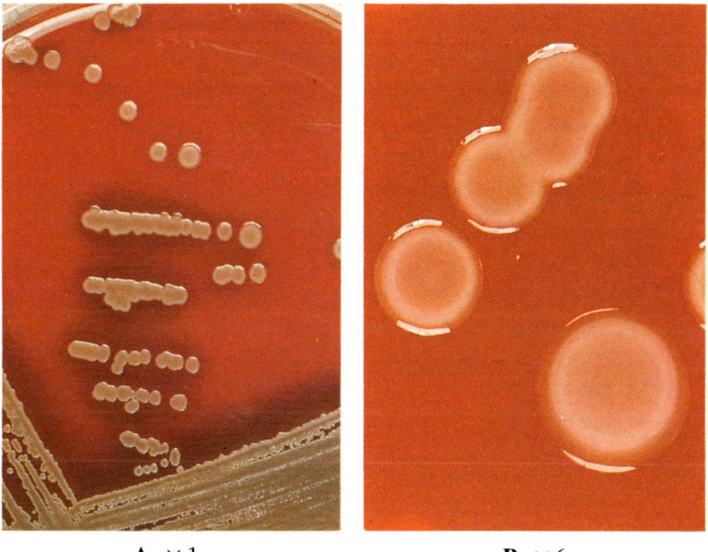

A, ×1.

B, ×6.

Figure 70 ***B. pumilus*** **NCTC 10327.** 24 hours; 37°C.

71

A

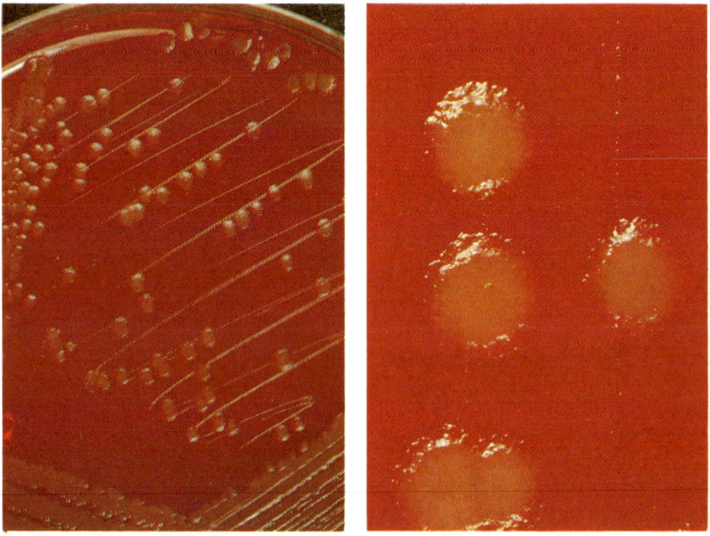

B

A, ×1. B, ×6.

Figure 71 ***B. firmus*** **NCTC 10335/ATCC 14575 (Type strain).** 24 hours; 37°C.

Bacillus firmus

Gordon *et al* (1973) say *B. firmus* is 'dutifully included in most comparative studies of the species of the genus *Bacillus* . . . (but) . . . appears to be sparsely represented in nature and to have few properties that appeal to biochemists, geneticists, microscopists, and molecular biologists working with bacteria'. But they also suggest it may be a somewhat neglected species.

We also 'dutifully' include *B. firmus* in its allotted position in Morphological Group 1, but the small number of strains present in culture collections make it a difficult species to describe precisely.

The strains examined were similar in appearance – creamy-grey, haemolytic colonies, 2–3mm in diameter (24 hours; 37°C), slightly moist with raised elevation and a slightly crenated edge.

72

A B

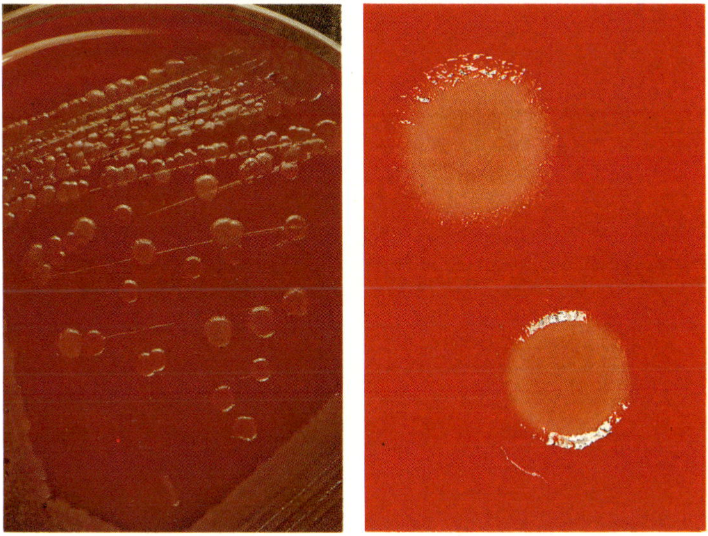

A, ×1. **B**, ×6.

Figure 72 ***B. firmus* NCIB 8162/ATCC 8247.** 24 hours; 37°C.

Cell size was approximately $0.7 \times 3\,\mu m$ and sporulating cells resembled those of *B. subtilis* (**Figure 64**) and *B. licheniformis* (**Figure 52**).

As seen in **Figure 1**, *B. firmus* stands alone in our primary identification scheme in being citrate negative and unable to grow anaerobically. It is also V-P negative which is relatively unusual in the Morphological Group 1 species (**Table 2**). Apart from this V-P reaction, *B. firmus* closely resembles the members of the *B. subtilis* group (Gordon, 1975).

The variant colonies seen in **Figure 72** arose consistently upon repeated subculture of single colonies.

Bacillus coagulans

Although as a group, members of the species *B. coagulans* are, in general, readily distinguishable biochemically from other *Bacillus* species (**Figure 1**; **Tables 1** and **2**), two distinct morphological types are identifiable – those whose spores do not swell their sporangia and those whose spores do swell their sporangia. As these are the distinguishing characteristics of Morphological Groups 1 and 2 respectively, *B. coagulans* is regarded as an intermediate species between Morphological Groups 1 and 2 (Gordon, 1975; Wolf and Sharp, 1981).

Cell types are also differentiable into those which resemble the bacterial cells of the subtilis group and those which are long and slender resembling the cells of *B. circulans* (Morphological Group 2 – Gordon *et al*, 1973).

Biochemical reactions are, in fact, similar to those of *B. licheniformis* with the exceptions of failure to hydrolyse gelatin and inability to grow in 7% salt broth.

B. coagulans is also a facultative thermophile growing well at 45–55°C with many strains also able to grow at 60°C (Wolf and Sharp, 1981). It is aciduric – that is, able to grow at relatively low pHs of down to 4.2 (Ingram, 1969).

Colonies at 24 hours (37°C) ranged from 1.5mm to 5mm. The majority of strains examined produced entire, raised colonies, slightly moist and butyrous, usually creamy-buff in colour and haemolytic. Strain NCTC 3992 (**Figure 75**) illustrates that there are exceptions to this colony description.

73

A

B

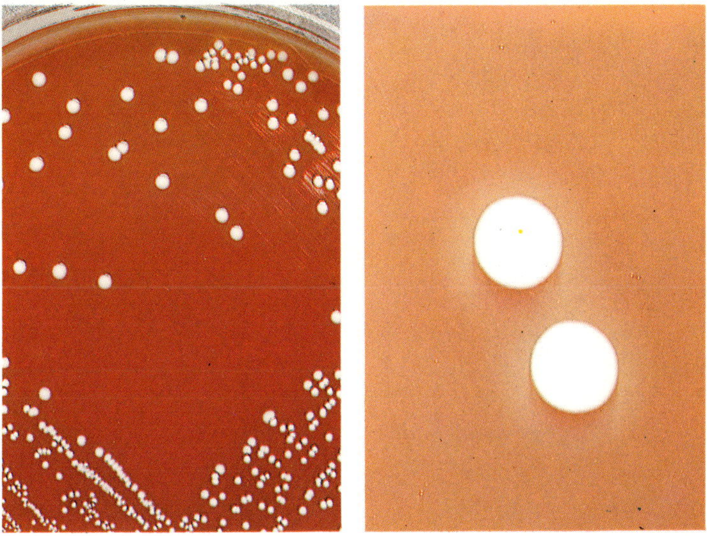

Figure 73 ***B. coagulans*** **NCTC 10334/ATCC 7050 (Type strain).** 24 hours; 37°C.

A, ×1. **B**, ×6. An overexposed photograph was chosen as showing better the colony shape, lack of surface detail and zones of haemolysis.

74

A

B

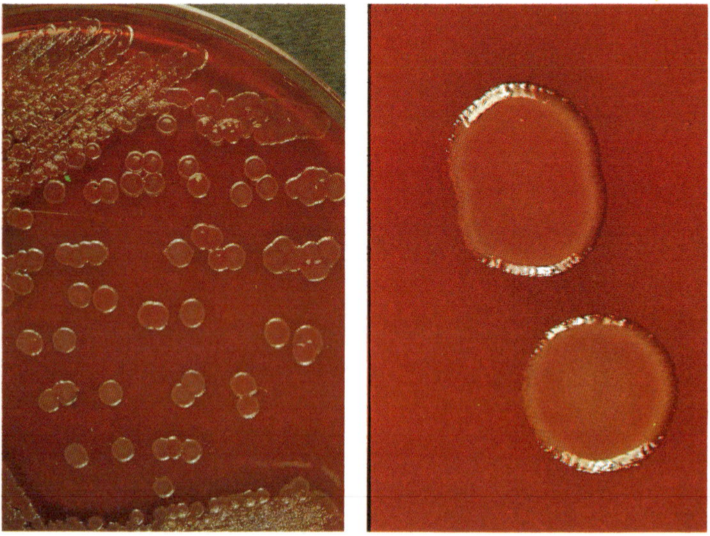

A, ×1. **B**, ×6.

Figure 74 ***B. coagulans*** **NCTC 3991.** 24 hours; 37°C.

75

A

B

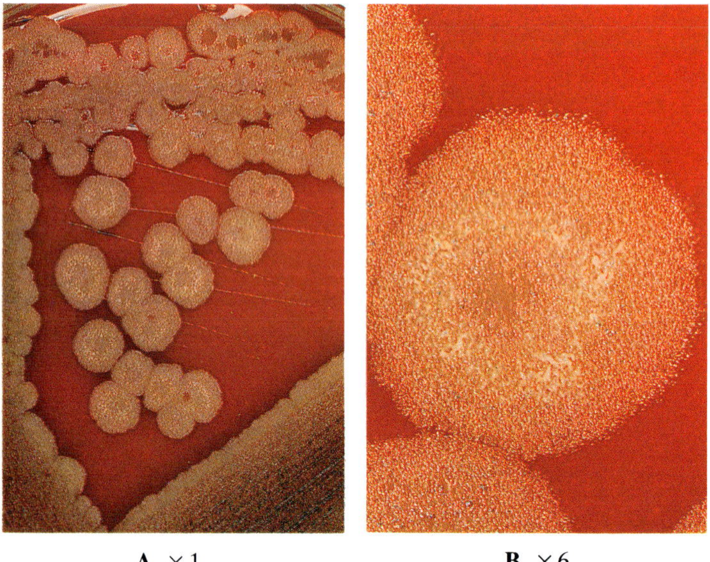

A, ×1. **B**, ×6.

Figure 75 ***B. coagulans*** **NCTC 3992.** 24 hours; 37°C.

The limited data available on DNA reassociation amongst *B. coagulans* strains support the concept that, despite the variable appearance (shape and size) of the bacterial cells and whether or not they are swollen by their spores, the species is, in fact, a homogeneous one (Priest, 1981).

Only one report of *B. coagulans* causing an infection (corneal abscess – van Bijsterveld and Richards, 1965) has come to our notice but the organism is isolated periodically from foods and pharmaceutical products.

Due in large part to its thermophilic properties and also to some extent to its ability to grow at relatively low pHs, it is one of the principal causes of 'flat-sour' spoilage (i.e. producing souring but no gas) in canned foods, including some acid foods such as canned fruits (Ingram, 1969; Norris *et al*, 1981).

It has also been isolated from medicated creams (Gilbert *et al*, 1981) and was the species most frequently encountered (52 of 77 samples) in an examination of proprietary liquid antacids (Willemse–Collinet *et al*, 1981).

An assay method for folic acid and commercial production of an enzyme for yeast cell lysis have utilized strains of *B. coagulans* (ATCC Catalogue, 1982).

Morphological Group 2

Defining characteristics

Sporangia (bacterial cells) definitely swollen by the spores.
Spores oval, rarely cylindrical, central, subterminal or terminal.
Gram-positive/negative/variable.

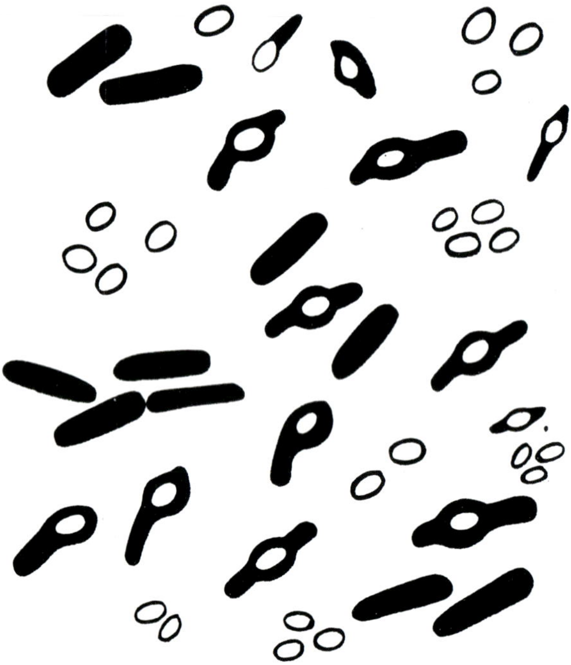

Group 2 species

1. *B. polymyxa*
2. *B. macerans*
3. *B. circulans*
4. *B. stearothermophilus*
5. *B. alvei*
6. *B. laterosporus*
7. *B. brevis*
8. 'B. pulvifaciens'
9. *B. popilliae*
10. *B. larvae*

76

A

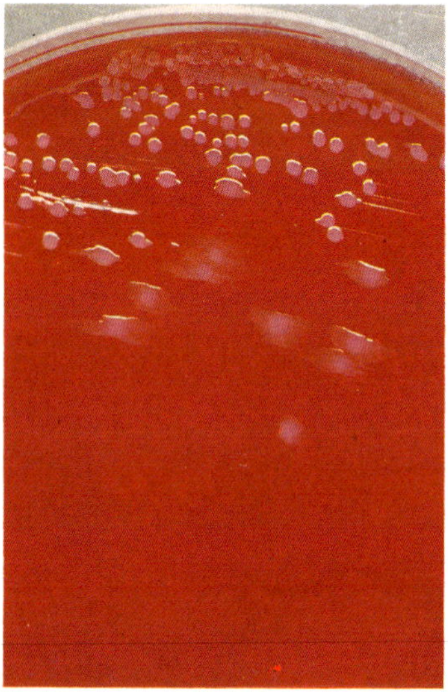

B

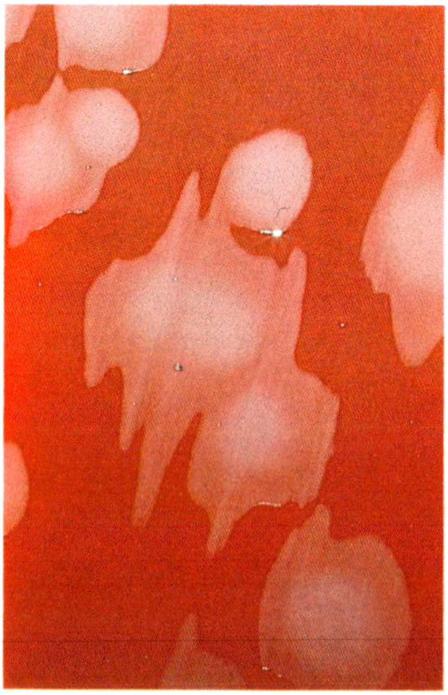

A, ×1. **B**, ×6.

Figure 76 ***B. polymyxa*** **NCTC 10343/ATCC 842 (Type strain).** 24 hours; 37°C.

Bacillus polymyxa

The name of this species derives from the Greek *poly* = much and *myxa* = slime or mucus and it was clearly so-called because of the frequently moist or mucoid appearances of the colonies of many strains (**Figures 76**, **78** and **80**).

The species is generally accepted as being easily recognized because little variability has been noted among different strains in their test characteristics (**Figure 1**; **Tables 1** and **3**). However, Priest *et al* (1981) and Logan and Berkeley (1981) have pointed out the close relationship between this species and *B. macerans* and *B. circulans*. These three species form a closely related group somewhat analagous to the 'subtilis group' (page 113).

77

A

B

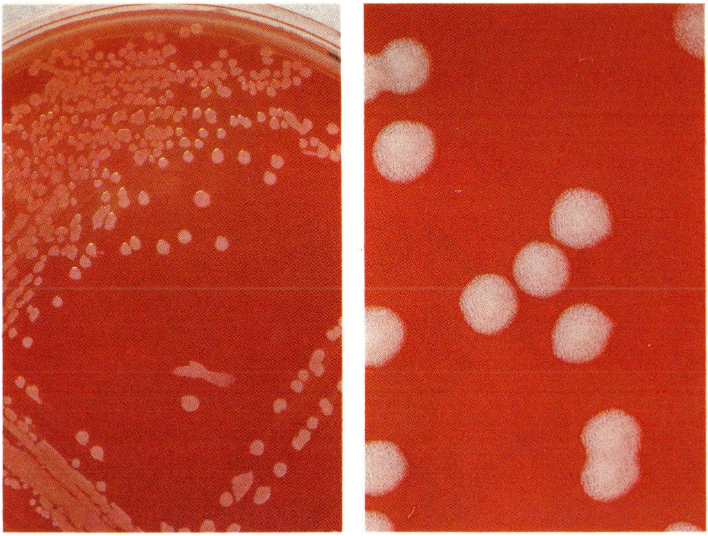

A, ×1. **B**, ×6.

Figure 77 ***B. polymyxa*** **NCTC 10343/ATCC 842 (Type strain).** 24 hours; 37°C.

The same strain as in **Figure 76** has been plated on to highly dried blood agar; the 'amoeboid' spreading of the colonies has been largely inhibited.

The colonies generally measure 2–4 mm in diameter, although some spread extensively **(Figure 78)** while others **(Figure 79)** may only produce relatively small colonies.

Gibson and Gordon (1974) describe the colonies as often having 'amoeboid spreading'. This is readily apparent in **Figure 76**, but as is shown in **Figure 77**, the extent to which this occurs is influenced by the degree of moistness of the agar.

78

A

B

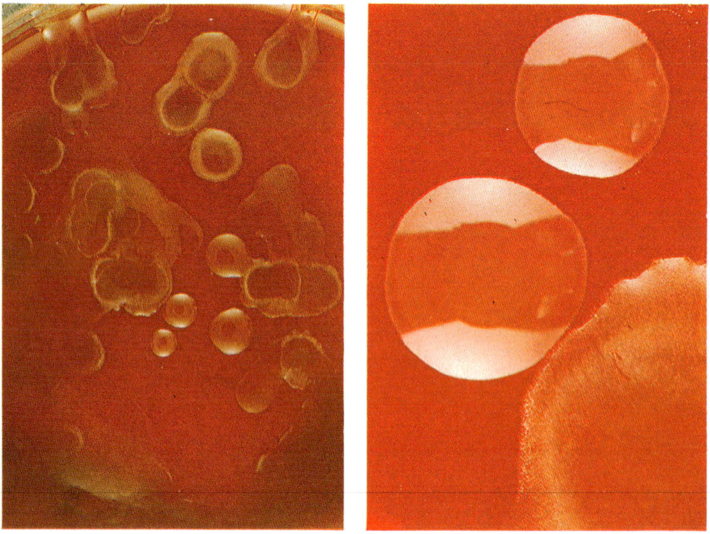

A, ×1. **B**, ×6.

Figure 78 ***B. polymyxa*** **F4373/78.** 24 hours; 37°C.

79

A

B

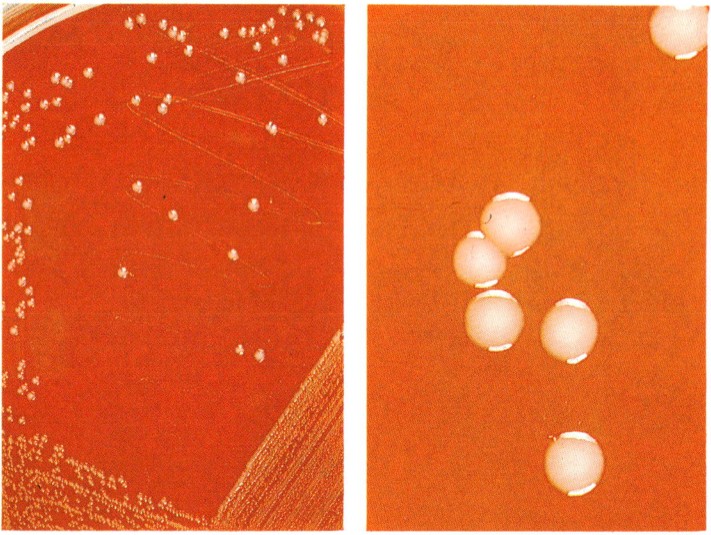

A, ×1. **B**, ×6.

Figure 79 ***B. polymyxa*** **NCTC 4747.** 24 hours; 37°C.

80

A B

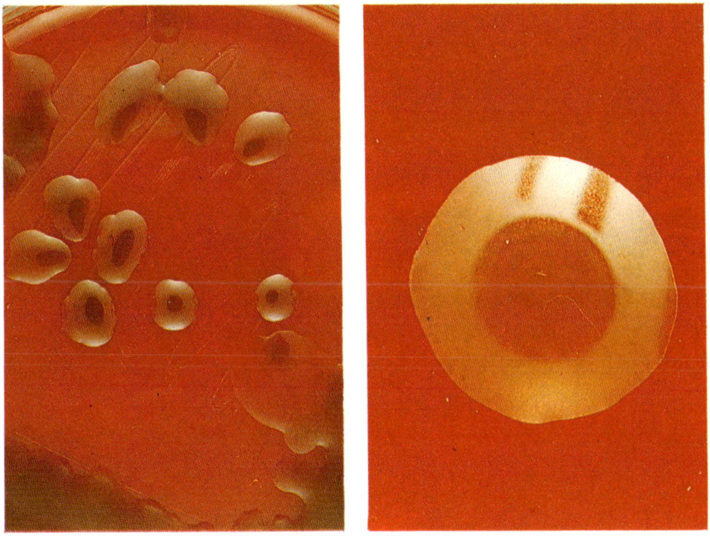

A, ×1. **B**, ×6.

Figure 80 ***B. polymyxa* F2789/78.** 24 hours; 37°C.

The 'mucus' of the mucoidal strains such as those seen in **Figures 78** and **80** dries out rapidly to give a crenated, wrinkled type of colony like those in **Figure 81**.

All strains produced a haemolysin although this was not always observed in young (24 hour) cultures.

The antibiotic polymyxin is a metabolite of certain strains of *B. polymyxa*. Strain NCTC 4747 (**Figure 79**), isolated from water by the Chicago Board of Health, featured in the first description of the antibiotic (Ainsworth *et al*, 1947) at the Wellcome Research Laboratories. At that time, a synonym for *B. polymyxa* was *B. aerosporus* – hence the trade name 'Aerosporin' (Burroughs–Wellcome Ltd, London).

There are 5 polymyxin types – A to E – all about equally effective against Gram-negative bacteria, but types B and E are the least toxic to the patient. Aerosporin is polymyxin B sulphate. Colistin is polymyxin E derived from '*B. polymyxa* var. *colistinus*' isolated in 1946 from the soil in the Fukushima Prefecture, Japan (Banyu Pharmaceutical Co. Ltd, Tokyo).

81

A

B

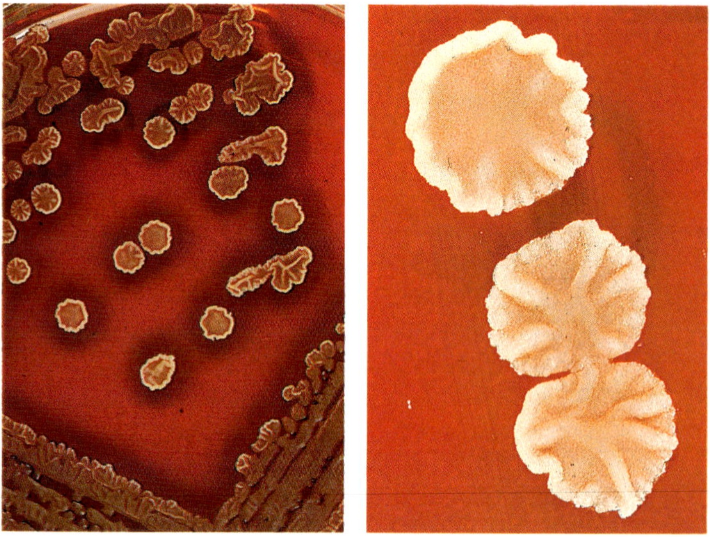

A, ×1. **B**, ×6.

Figure 81 ***B. polymyxa* F4686/78.** 24 hours; 37°C.

Strains of *B. polymyxa* are used in the production of other biochemicals as well as polymyxin antibiotics.

No reports of infections due to *B. polymyxa* have come to our notice. The species is not infrequently isolated from foods and, having a tolerance to relatively low pHs, it occasionally causes spoilage in 'acid' (pH 3.7–4.5) foods such as canned fruits (Ingram, 1969).

B. polymyxa was also among the *Bacillus* species found in medicated creams (Gilbert *et al*, 1981) and proprietary antacids (Willemse–Collinet *et al*, 1981).

The bacterial cell size averages $0.7 \times 3.5\,\mu m$. Spores are oval, being centrally, subterminally or terminally placed, and markedly swelling the sporangium. In **Figure 83**, it is readily seen that the width of the spores greatly exceeds that of the bacterial rods.

82

A 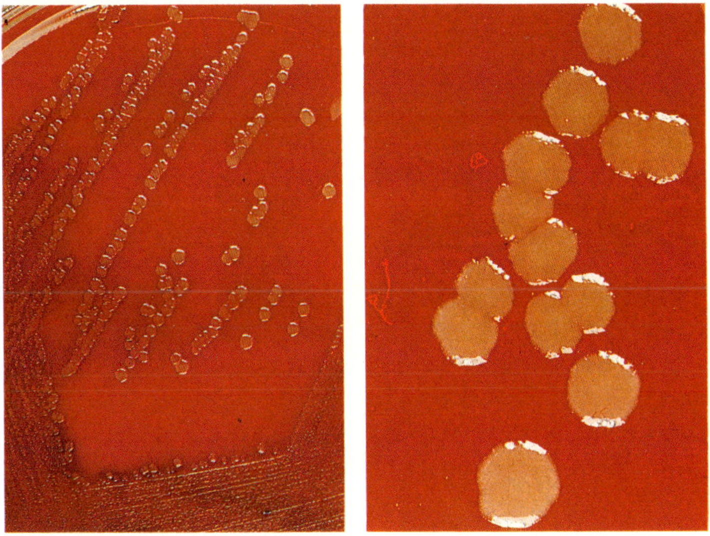 B

A, ×1. B, ×6.

Figure 82 ***B. polymyxa*** **NCTC 7575.** 24 hours; 37°C.

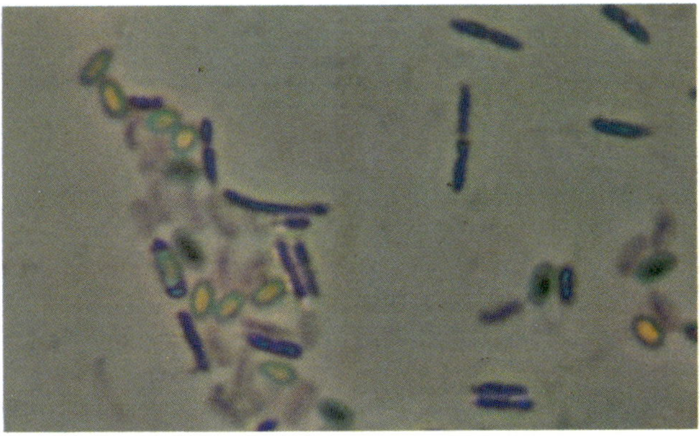

83

Figure 83 Spores and vegetative cells of *B. polymyxa*. Phase contrast; ×1600.

Bacillus macerans

The close relatedness between *B. macerans*, *B. polymyxa* and *B. circulans* was pointed out on page 138, but apart from colonial appearance, *B. macerans* can be differentiated from *B. polymyxa* by its inability to hydrolyse casein and its negative V-P reaction (**Table 3**). It is more difficult to distinguish *B. macerans* and *B. circulans*; the single reliable criterion accepted by Gordon *et al* (1973) was production of gas in ammonium salt sugars (*B. macerans* = positive; *B. circulans* = negative).

Typical colonies measure 2–3mm in diameter, are creamy-grey in colour and produce slight haemolysis. The colony has a low convex shape with an entire or irregular edge and a fine granular or amorphous surface.

Cell and spore sizes and the position of the spore in the sporangium are similar to those described for *B. polymyxa* (**Figure 83**).

We know of just one report of *B. macerans* infection – an infected excision site of a malignant melanoma (Ihde and Armstrong, 1973).

The organism is periodically isolated from foodstuffs and like *B. polymyxa*, is able to grow at relatively low pHs and is therefore occasionally a spoilage organism in fairly acid foods.

The variant colonies seen in **Figure 84** arose consistently upon repeated single colony subculturing.

84

A

B

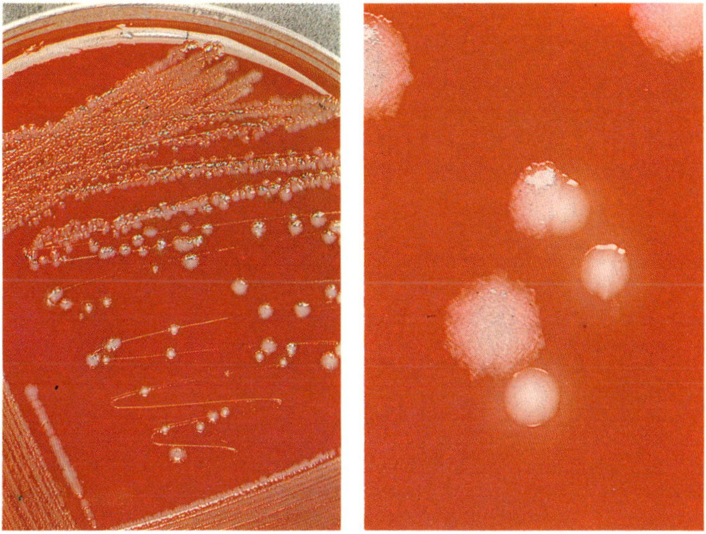

A, ×1. **B**, ×6.

Figure 84 ***B. macerans*** **NCTC 6355/ATCC 8244 (Type strain).** 24 hours; 37°C.

85

A

B

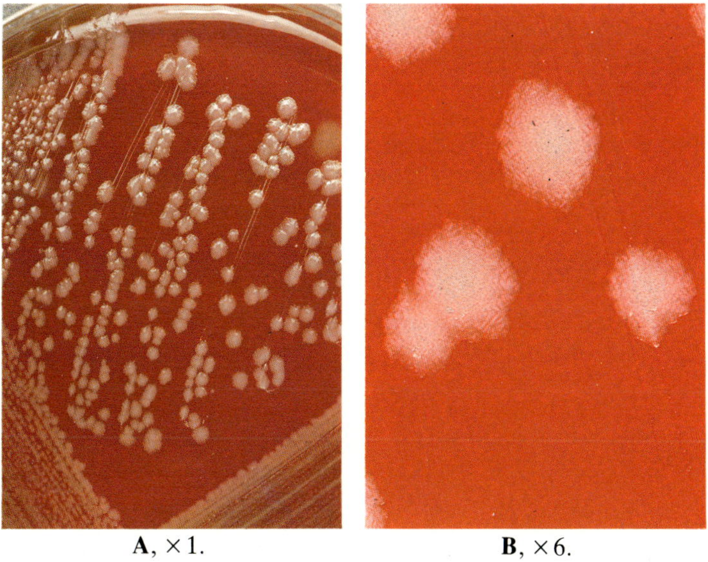

A, ×1. **B**, ×6.

Figure 85 ***B. macerans*** **NCTC 7588.** 24 hours; 37°C.

86

A

B

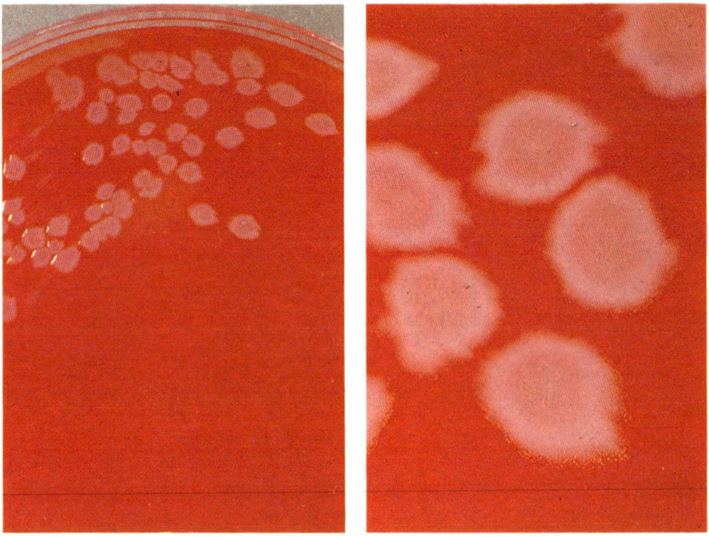

A, ×1. B, ×6.

Figure 86 ***B. circulans*** **NCTC 2610/ATCC 4513 (Type strain).**
24 hours; 37°C.

Bacillus circulans

B. circulans got its name in the last century when Jordan (1890) noted that the interior of its colonies, when viewed under the microscope had a circular motion like that of protoplasm. Strains have also been observed to produce microcolonies which rotate over the surface of agar (Gordon *et al*, 1973) although we did not observe this with any of the strains we examined.

As indicated in the number of variable characteristics apparent in **Table 3**, the species is something of a heterogeneous one and it is often difficult to distinguish it from *B. macerans*. Spore morphologies of *B. circulans* and *B. macerans* under the electron microscope are markedly different (Bradley and Franklin, 1958), but the best differentiation in the routine laboratory is production of gas in carbohydrate media by *B. macerans*; *B. circulans* is anaerogenic.

87

A

B

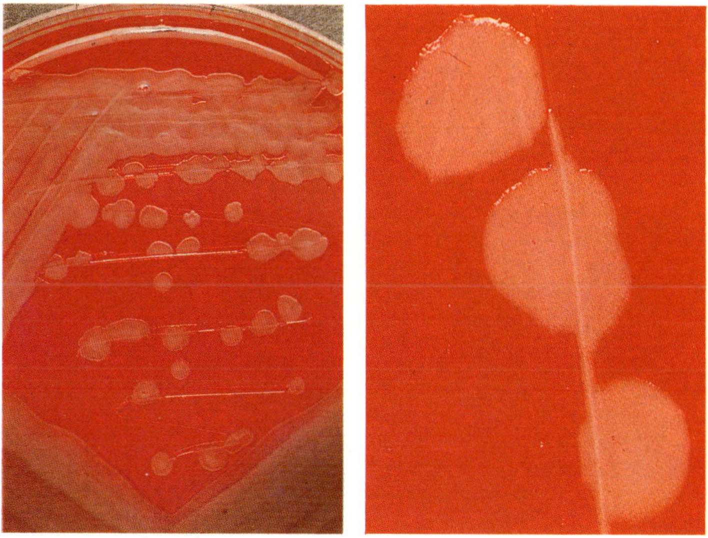

A, ×1. **B**, ×6.

Figure 87 ***B. circulans*** **NCTC 9432.** 24 hours; 37°C.

The heterogeneity of the species has been confirmed by studies on DNA G + C contents and reassociation and by extensive studies using 119 API system tests (Gibson and Gordon, 1974; Logan and Berkeley, 1981; Priest *et al*, 1981).

Colony size ranges from 2–4 mm; colonies have entire or slightly irregular edges, effuse (thin and widely spread) to convex elevation with a smooth translucent surface and a butyrous consistency. Colony colour ranges from creamy-grey to buff.

All strains we examined were haemolytic.

The bacterial cell size averages $0.6 \times 4\,\mu m$ and contains oval spores, mostly either subterminal or terminal and which swell the sporangia.

88

A

B

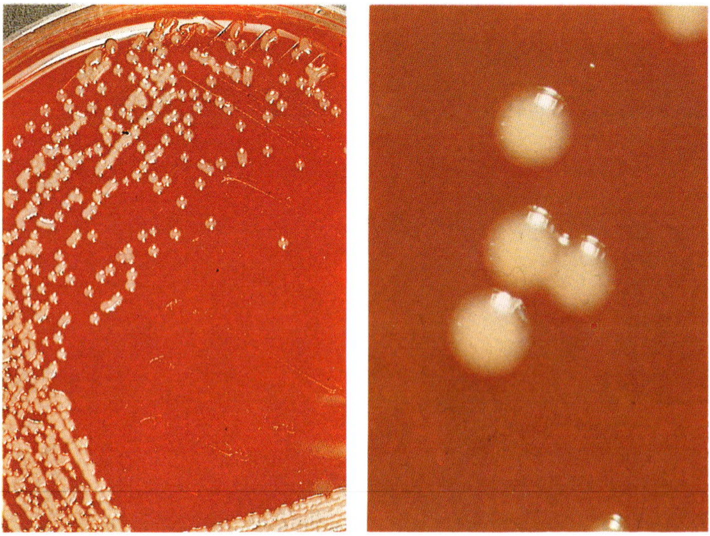

A, ×1. **B**, ×6.

Figure 88 ***Bacillus* sp. NCTC 4821.** 24 hours; 37°C.

Formerly named *B. carotarum*, we found this culture to be a strain of *B. circulans*.

A case of meningitis in a child due to *B. circulans* has been recorded (Boyette and Rights, 1952) and strains of the species have been associated with spoilage of canned cured meats (Norris *et al*, 1981). The organism was among other *Bacillus* species found contaminating proprietary liquid antacids (Willemse–Collinet *et al*, 1981).

On the positive side, a number of strains are listed (ATCC Catalogue, 1982) as being used in the production of certain antibiotics, enzymes and other biochemical agents and strain ATCC 9966 (not shown) is a standard strain for streptomycin assays.

The inclusion of *B. circulans* together with *B. subtilis* at a certain stage in the processing of cigar tobacco apparently enhances the aroma of the final product (English *et al*, 1967).

89

A B

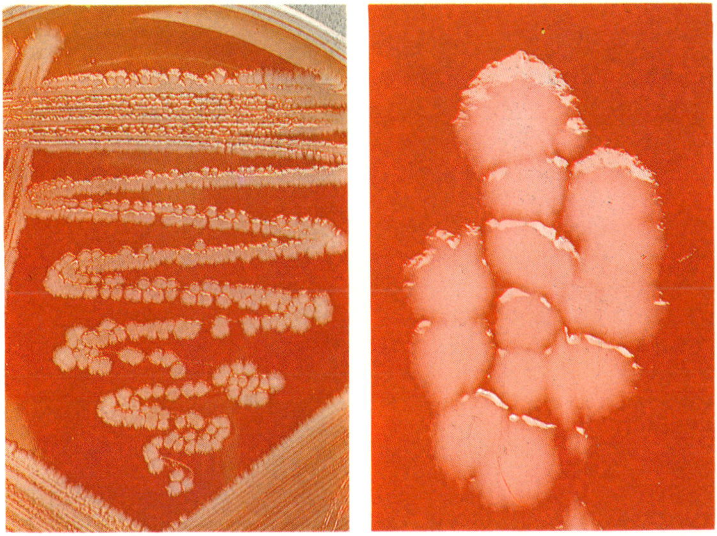

A, ×1. **B**, ×6.

Figure 89 ***B. circulans*** **NCTC 5846.** 24 hours; 37°C.

90

A B

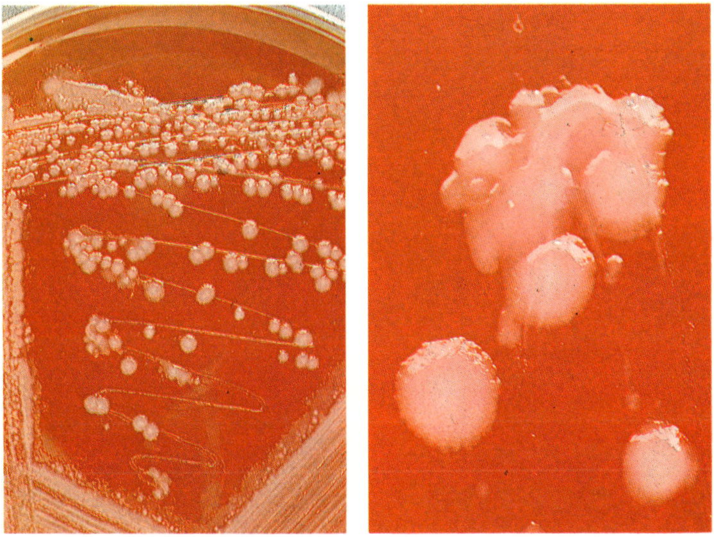

A, ×1. **B**, ×6.

Figure 90 ***B. circulans*** **NCTC 5895.** 24 hours; 37°C.

Bacillus stearothermophilus

B. stearothermophilus is one of nature's most interesting organisms; it is an obligate thermophile with an optimal vegetative cell growth temperature of 55–60°C and maximum growth temperature usually in the order of 65°C. Some strains, however, have been known to be capable of growth at as high as 75°C – a temperature at which denaturation of proteins and enzyme inactivation in, and destruction of other organisms readily occurs. Only a few strains are capable of growth below 40°C – near the upper limit for many bacteria!

As reviewed by Norris *et al* (1981), as well as being found in desert sand, soils in the tropics and water of hot springs, viable spores have been found in Arctic soils and waters and even deep cores of ocean basins estimated to be thousands of years old.

Hand in hand with the high temperature preference of the vegetative cells is an even higher heat-resistance among the spores of *B. stearothermophilus* as compared with those of other *Bacillus* species and as a result of this, the species has long been of importance and concern as a cause of flat-sour (acid production but no gas) spoilage in the canned food and dairy industries; in fact, the first named and described of the species were isolated from cans of spoiled maize (corn) and string beans (Gordon *et al*, 1973).

B. stearothermophilus is something of a taxonomic enigma and it has been considered by some that the species merely represents thermophilic variants of other mesophilic *Bacillus* species, such as *B. circulans*, *B. megaterium* or *B. subtilis*. It has clearly been accepted now by taxonomists as a species in its own right although all are agreed that it is a heterogeneous one composed of at least three distinct groups. The details of this are beyond the scope of this atlas; it is a complex topic which is extensively reviewed by Wolf and Sharp (1981).

91

A

B

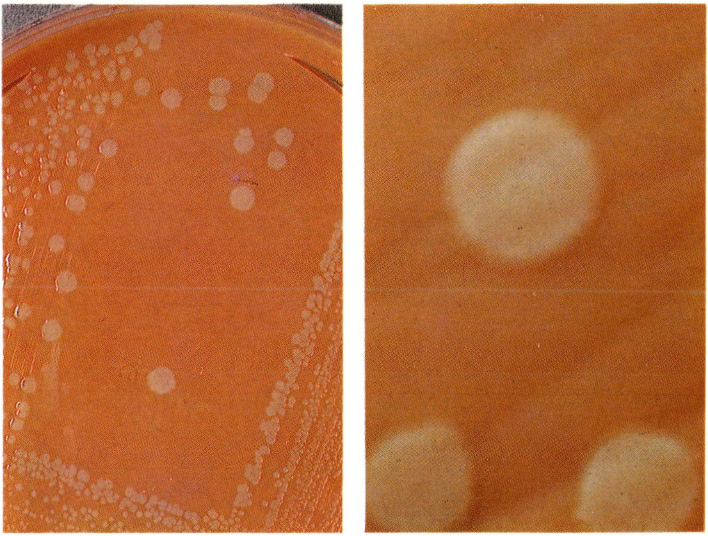

A, ×1. **B**, ×6.

Figure 91 ***B. stearothermophilus*** **NCTC 10339/ATCC 12980 (Type strain).** 24 hours; 60°C.

Note the blood agar has 'chocolatized', dried and split under the incubation conditions.

Although the spores can survive extremes of cold, if placed in a rich nutritional environment at an ambient temperature of around 30°C, they will vegetate but the vegetative organisms will die at this temperature. The phenomenon can occur in canned foods, for example, and is termed 'auto-sterilization'. For this reason, it is often unnecessary to give canned foods the severe heat treatments that would be necessary to eliminate these thermophilic spores – and which would also ruin the foods. From this arises the term 'commercial sterility' in which viable thermophilic spores survive the processing of the food, but under all reasonable storage conditions, the pack behaves for all practical purposes as if sterile (Ingram, 1969). The vegetative cells are unable to form new spores at temperatures as low as 30°C.

92

A

B

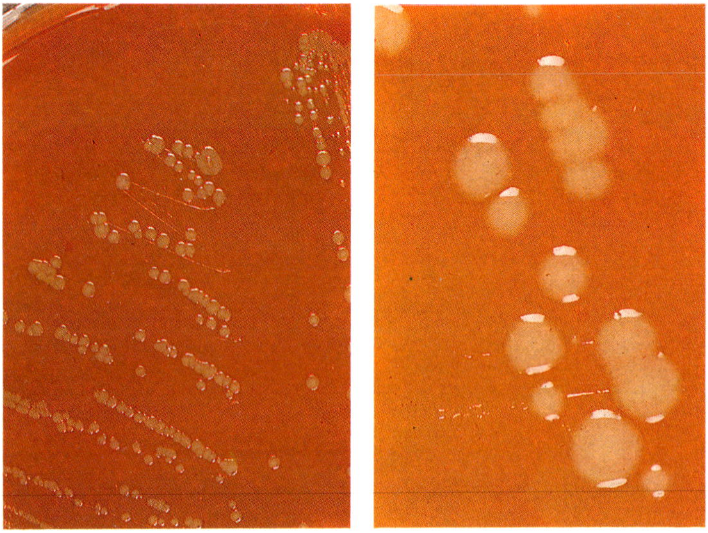

A, ×1. **B**, ×6.

Figure 92 ***B. stearothermophilus*** **NCTC 10007/ATCC 7953.** 24 hours; 60°C.

The extreme heat resistance of *B. stearothermophilus* spores has been utilized in a number of situations to determine the efficiency of heat sterilization procedures. Strain NCTC 10007/ATCC 7953 (**Figure 92**) is a standard strain for heat sterilization control.

The organism also finds use in assays of antibacterial drugs in patient serum and antibiotic residues in milk and dairy products. The advantages of using *B. stearothermophilus* in these types of test are (1) its sensitivity to a wide range of antibiotics, (2) the rapidity with which the organism grows at its high temperature of incubation (63–66°C) in these assays and (3) in the case of milk and dairy products, the lack of interference in the test by other organisms in the milk at this high incubation temperature.

In the case of testing for antibiotic residues in milk, the standard strains used, referred to in literature on the subject as *B. stearothermophilus* var. *calidolactis* (strain ATCC 10149 or the Netherlands Institute for Dairy Research strain C953) are sensitive to essentially all the antibiotics used in treatment of bovine mastitis or other bovine diseases and also to non-antibiotic chemical inhibitors likely to be found in milk (van Os *et al*, 1975; Ouderkirk, 1979). Commercial test kits consisting of solid medium seeded with *B. stearothermophilus* var. *calidolactis* are available (see Appendix 3).

After 24 hours at 60°C, typical colonies measured 1.5–3 mm in diameter and had an entire circular edge, convex elevation, a smooth translucent surface and a butyrous consistency.

Haemolysis was apparent in the form of a slight greening of the agar around the colonies.

Bacterial rod sizes average $0.8 \times 2.8\,\mu m$. Spores are oval, subterminal or terminal and with some exceptions (Wolf and Sharp, 1981) markedly swell the sporangium.

93

A 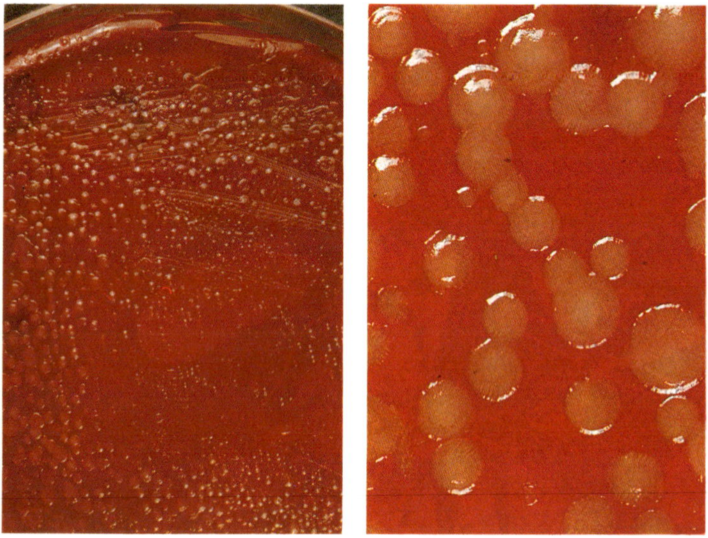 B

A, ×1. B, ×6.

Figure 93 ***B. alvei*** **NCTC 6352/ATCC 6344 (Type strain).** 24 hours; 37°C.

Bacillus alvei

This species was first isolated in 1885 and is one of the earliest named *Bacillus* species. It was isolated from diseased bees and its name (Latin *alvei* = from a beehive) derives from this.

Other strains of *B. alvei* were subsequently isolated from diseased bees also and it was thought at first to be an insect pathogen but this was later discounted (reviewed by Gordon *et al*, 1973).

It has, however, apparently been isolated from patients in a small number of cases although the details of the cases have not been published (Gordon *et al*, 1973).

94

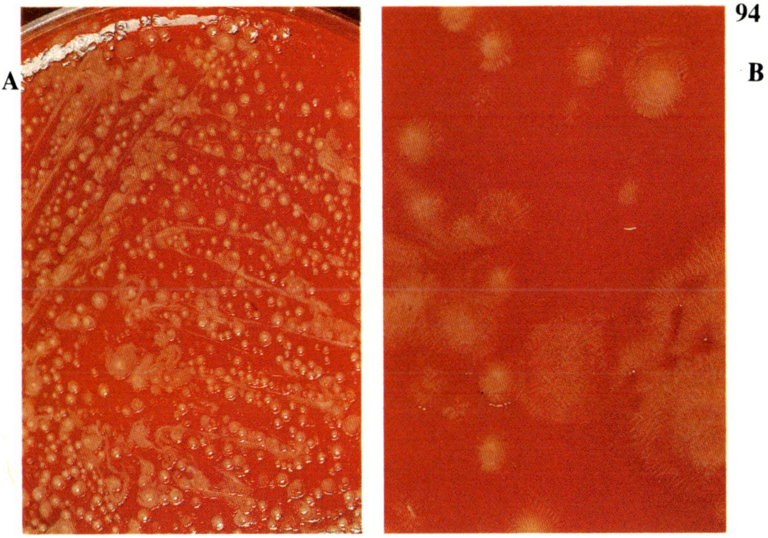

A, ×1. B, ×6.

Figure 94 ***B. alvei*** **NCTC 3349.** 24 hours; 37°C.

The organism is actively motile and, when very highly motile, results in spreading colonies and 'motile microcolonies'. Consequently, after incubation, growth is found all over the plate and not just along the streak-line of the inoculating loop (**Figures 93** and **94**).

When discrete, the colonies range from 2–4 mm in diameter (24 hours; 37°C), are circular or irregular, low convex to convex and have a smooth glistening, translucent or opaque creamy-grey surface.

B. alvei is distinguished from *B. circulans*, the other species capable of forming motile microcolonies, principally by indole, V-P and ammonium salt sugar reactions (**Table 3**) although 'intermediate' results may be obtained in V-P and ammonium salt sugar reactions with some strains of *B. circulans* (Gibson and Gordon, 1974).

In contrast to other members of Morphological Group 2, *B. alvei* and *B. polymyxa* are usually V-P positive; these two are differentiated by nitrate reduction and indole production.

The bacterial rod size averages $0.5 \times 3.4\,\mu m$. Spores are oval, central, subterminal or terminal and swell the sporangium.

95

A

B

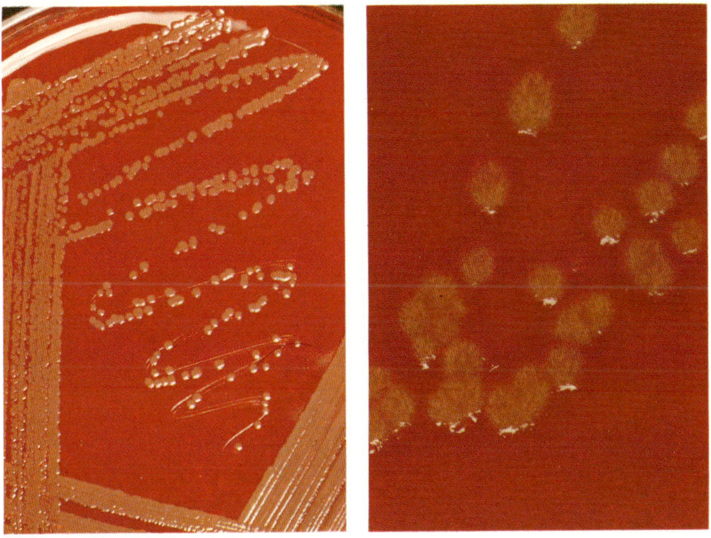

A, ×1. B, ×6.

Figure 95 ***B. alvei*** **NCTC 7583.** 24 hours; 37°C.

96

A

B

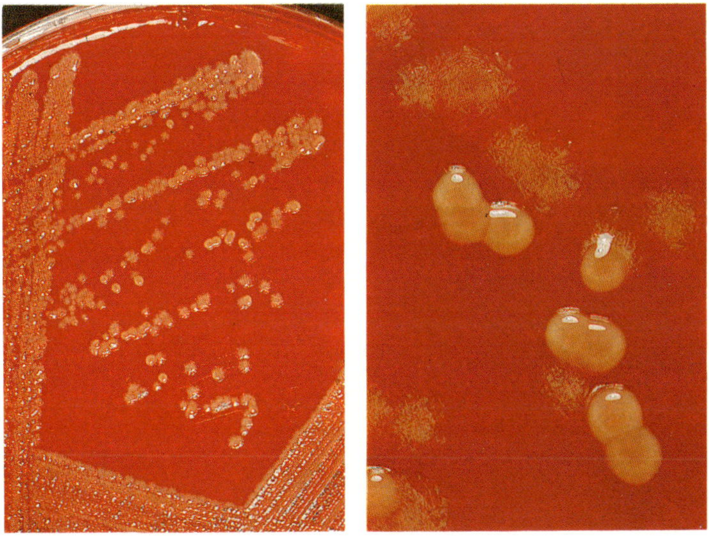

A, ×1. B, ×6.

Figure 96 ***B. alvei*** **NCTC 3324.** 24 hours; 37°C.

97

A

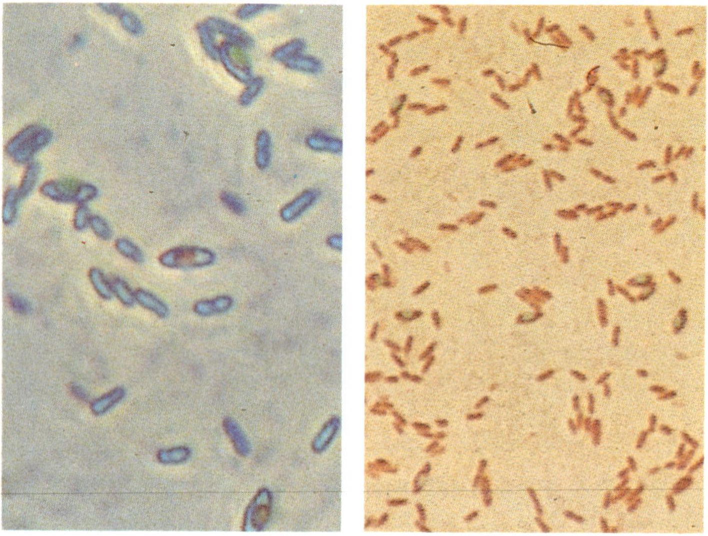

B

A, Phase contrast. ×1600. B, Malachite green stain. ×1300.

Figure 97 Vegetative cells and spores of *B. laterosporus*.

The magnifications given are calculated from the lens magnifications on the objectives and camera eyepieces. As can be seen, the effect of phase contrast has been to make the apparent magnification of the cells in **A** twice that of those in **B**.

Bacillus laterosporus

A glance at **Figure 97** reveals how this species came to be so-named. The important diagnostic feature is the canoe-like appearance of the sporulating cell resulting from the manner in which the spore lies laterally in the spindle-shaped sporangium (vegetative cell body). The spore is apparently cradled in a C-shaped parasporal body when mature and when released from the lysed sporangium. This makes the rim of the spore appear thicker on one side than the other.

A caution is given by Gibson and Gordon (1974) in that induction of sporulation is difficult in some strains and that lateral spores may occur in other species of *Bacillus*.

The vegetative cells are approximately $0.6 \times 3.5\,\mu m$.

98

A

B

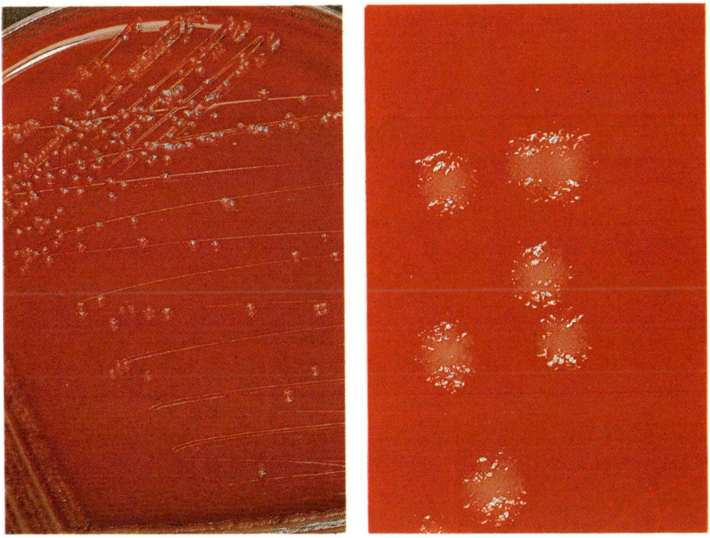

A, ×1. **B**, ×6.

Figure 98 ***B. laterosporus*** **NCTC 6357/ATCC 64 (Type strain).** 24 hours; 37°C.

B. laterosporus is not, apparently, a common *Bacillus* species. The first described strain was isolated in 1916 from water. Other strains have been isolated from water, soil and diseased bees although, like *B. alvei*, it is not thought that the organism is an insect pathogen. One isolation of *B. laterosporus* was recorded by Willemse–Collinet *et al* (1981) in their examination of proprietary liquid antacids.

As judged by DNA base determination studies, the species appears to be a homogeneous one and well demarcated from other species (Priest *et al*, 1981).

Colonies of the species we examined measured 2–4mm in diameter and were circular with a slightly irregular edge. Colony elevation was low convex to convex and the creamy-grey colonies had a smooth glistening surface with a deep opaque centre.

Slight haemolysis was apparent around the colonies.

Biochemically, *B. laterosporus* is the only Morphological Group 2 species to produce lecithinase; unlike Group 1 LV positive species, it ferments mannitol ammonium salt sugar.

99

A B

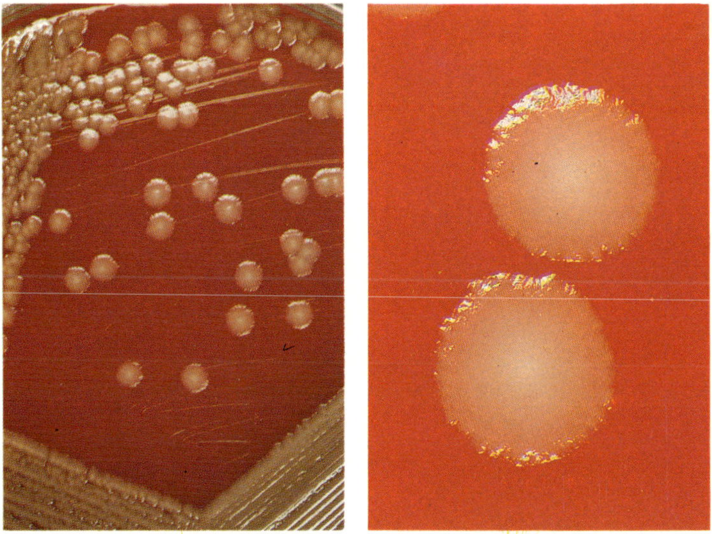

A, ×1. **B**, ×6.

Figure 99 ***B. laterosporus*** **NCTC 2613/ATCC 4517.** 24 hours; 37°C.

100

A B

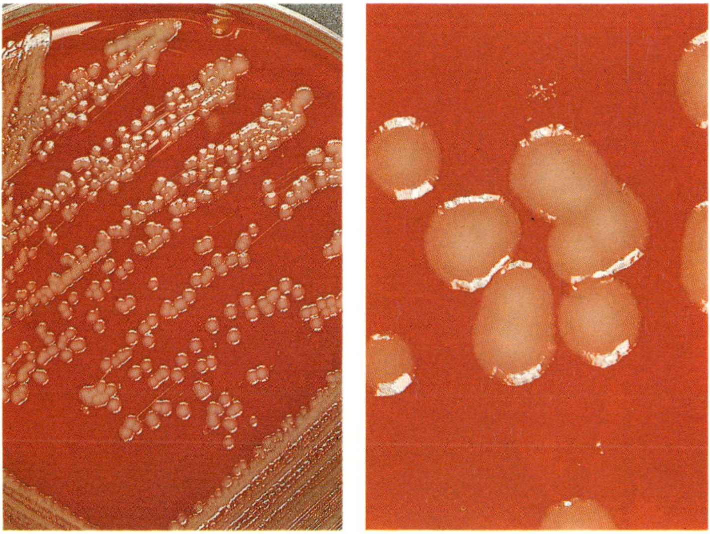

A, ×1. **B**, ×6.

Figure 100 ***B. laterosporus*** **NCTC 7579.** 24 hours; 37°C.

101

A

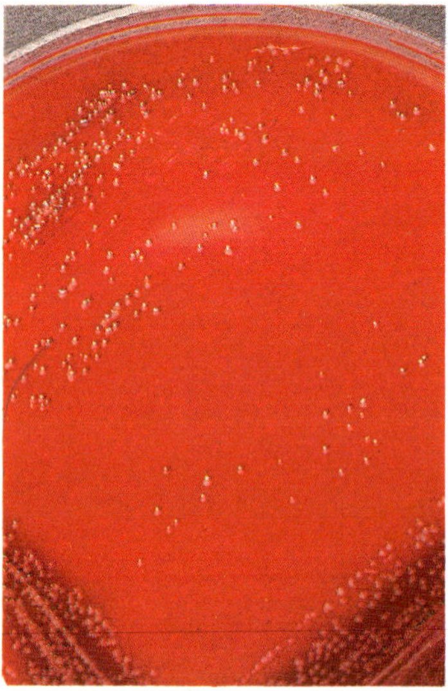

B

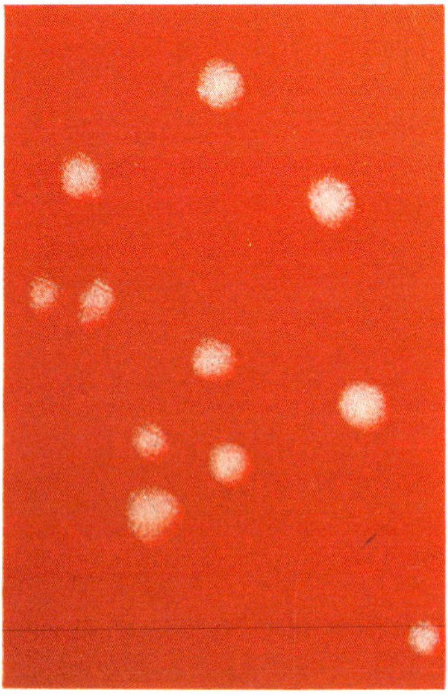

A, ×1. **B**, ×6.

Figure 101 ***B. brevis*** **NCTC 2611/ATCC 8246 (Type strain).** 24 hours; 37°C.

Bacillus brevis

It is not entirely clear yet whether *B. brevis* comprises a compact homogeneous species or whether it is differentiable into more than one subtype, at least by sophisticated methods of DNA base determination. Biochemically, it appears to be a discrete species (Gordon *et al*, 1973; Logan and Berkeley, 1981; Priest *et al*, 1981).

The colonies of the type strain, NCTC 2611/ATCC 8246 at 24 hours (37°C) were only of a pinpoint size (**Figure 101A**) although under magnification, they had a typical 'ASB' appearance (**Figure 101B**). This became more marked upon further incubation (**Figure 102**). The colonies of other strains examined were approximately 2mm in diameter at 24 hours.

A

102

B

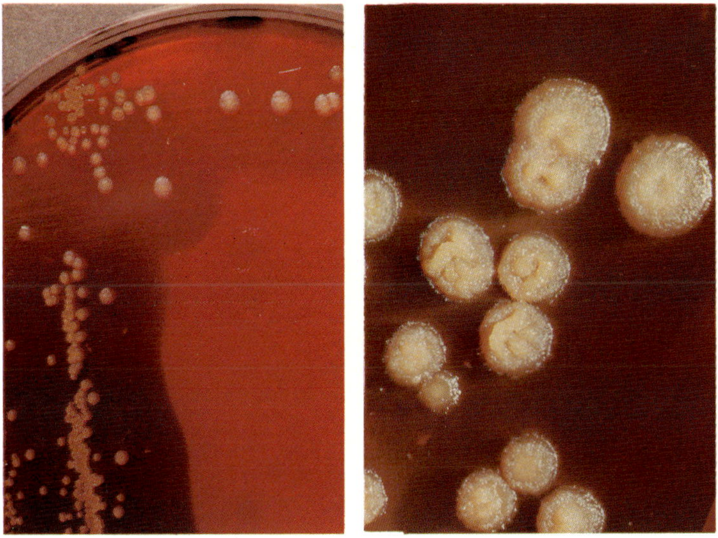

A, × 1. Note marked haemolysis. **B**, × 6.

Figure 102 ***B. brevis*** **NCTC 2611/ATCC 8246.** 72 hours; 37°C.

In general, colonies were low convex and circular with an irregular edge and granular surface.

Haemolysis was usually apparent at 24 hours although with NCTC 2611 it was not visible until 48 hours. The colonies tend to take up a greenish tinge as a result of the haemolysis they produce (see discussion under **Figure 33**).

The granular appearance and greenish tinge of the colonies may lead to an initial confusion with *B. cereus* but differentiation is readily established by lecithovitellin and other biochemical reactions.

103

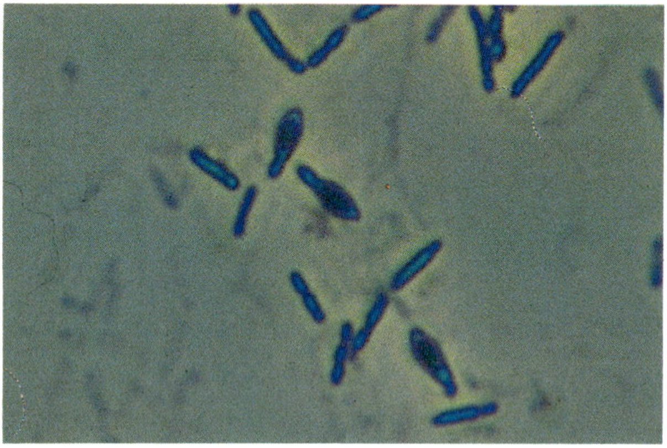

Figure 103 Vegetative cells and spores of *B. brevis*. Phase contrast. × 1600.

Recognition of the species begins with observation of the elliptical spores which distend the sporangia into spindle-shaped cells (**Figure 103**). The spore may be central, subterminal or terminal. The average vegetative cell size is $0.75 \times 2.5\,\mu m$.

Occasionally facultatively thermophilic strains are encountered able to grow at temperatures of up to 60°C, but this property is not common to all members of the species. It may, however, have bearing on the periodic isolation of *B. brevis* from processed foods as well as from foods in general. For example, Norris *et al* (1981) record a report of isolation from faulty cans of evaporated milk.

The organism has been isolated in large numbers from foods associated with four incidents of food poisoning (Gilbert *et al*, 1981). It was also isolated from several of the proprietary liquid antacids examined by Willemse–Collinet *et al* (1981).

We are aware of a single case report of *B. brevis* infection in which it was implicated together with an achromogenic pseudomonad in corneal ulceration (van Bijsterveld and Richards, 1965).

Several strains of the species elaborate the antibiotics gramicidin and tyrocidin. An unnamed antibiotic, 342–14–1 (US Pat. 3,969,501) is also produced by a strain of *B. brevis*.

104

A

B

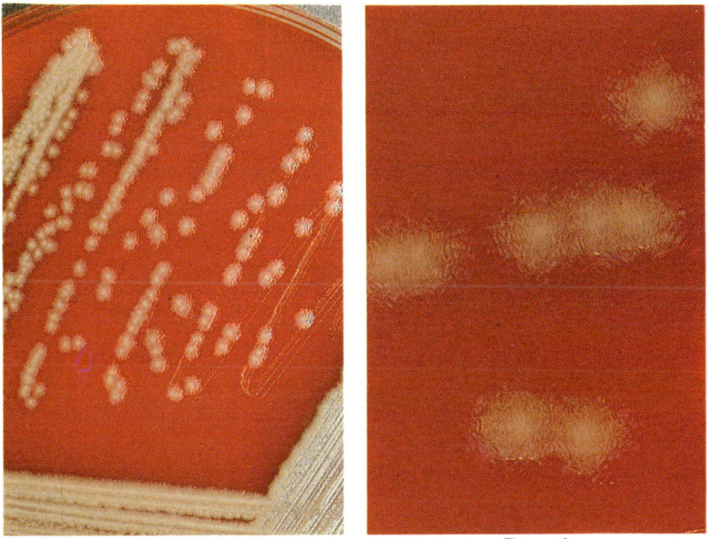

A, ×1. **B**, ×6.

Figure 104 ***B. brevis*** **NCTC 7096/ATCC 8246.** 24 hours; 37°C.

105

A

B

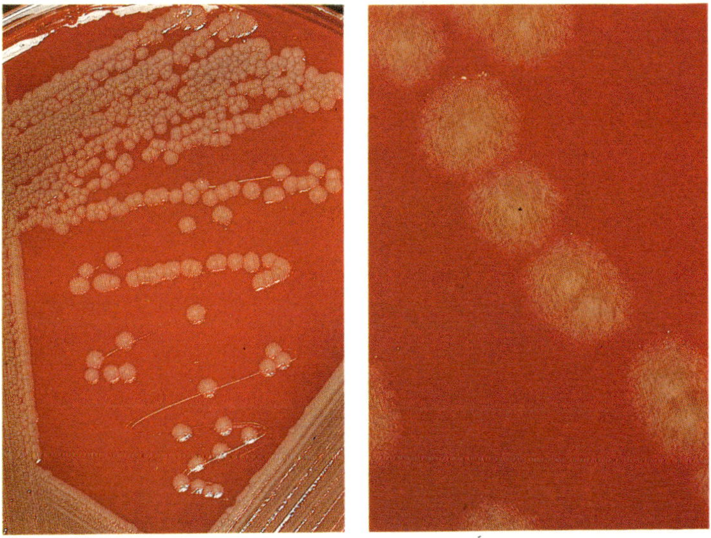

A, ×1. **B**, ×6.

Figure 105 ***B. brevis*** **NCTC 7577.** 24 hours; 37°C.

106

A

B

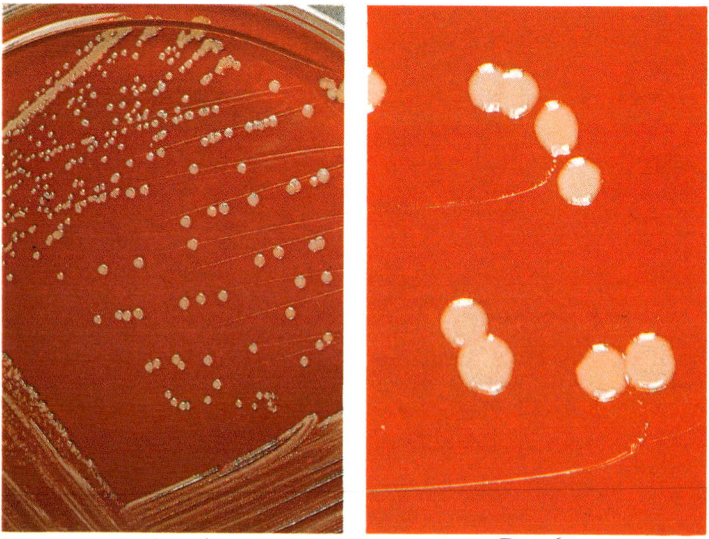

A, ×1. B, ×6.
Figure 106 'B. pulvifaciens' CCM 38. 24 hours; 37°C.

'Bacillus pulvifaciens'

Although culture collection strains with this species name are available, 'B. pulvifaciens' was not included in the Approved Lists of Bacterial Names (1980). Gordon *et al* (1973) supported by Priest *et al* (1981) felt that the group was most closely related to *B. larvae*, but as yet, the existing strains of 'B. pulvifaciens' have not been officially incorporated into the species *B. larvae*.

'B. pulvifaciens' strains give delayed weak-positive catalase reactions and, as such, are placed here following the concept of Gordon *et al* (1973) that they are a link between catalase-producing *Bacillus* (most species) and the catalase-negative insect pathogens *B. larvae* and *B. popilliae* of Morphological Group 2.

Although not definitely established as an insect pathogen, 'B. pulvifaciens' has been isolated from infected honey-bee larvae (Gordon *et al*, 1973), but there is little reason at present to believe it is pathogenic for mammals. It grows readily on blood agar at 37°C producing, in 24 hours, low convex, entire, smooth and glistening haemolytic colonies, 2–3mm in diameter and could, therefore, be encountered in the clinical laboratory.

The vegetative cells are small – approximately $0.4 \times 2.25\,\mu m$ – and produce central or subterminal oval spores which swell the sporangium. The species has not been encountered frequently.

107

A

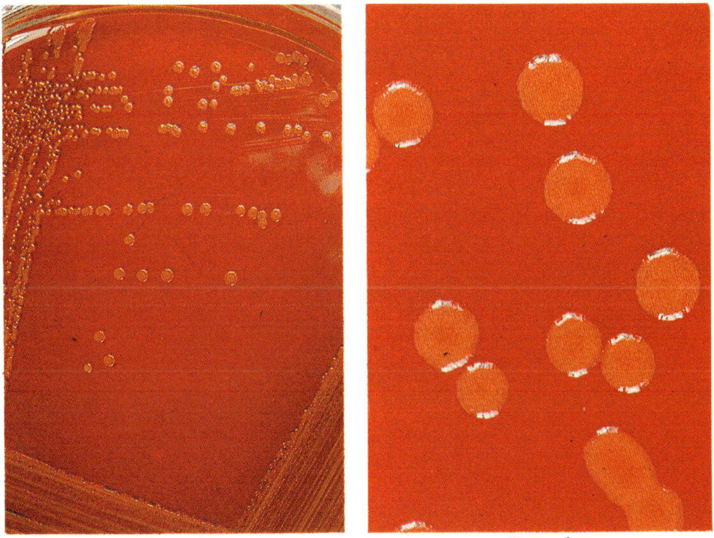

B

A, ×1. **B**, ×6.

Figure 107 'B. pulvifaciens' CCM 39. 24 hours; 37°C.

108

A

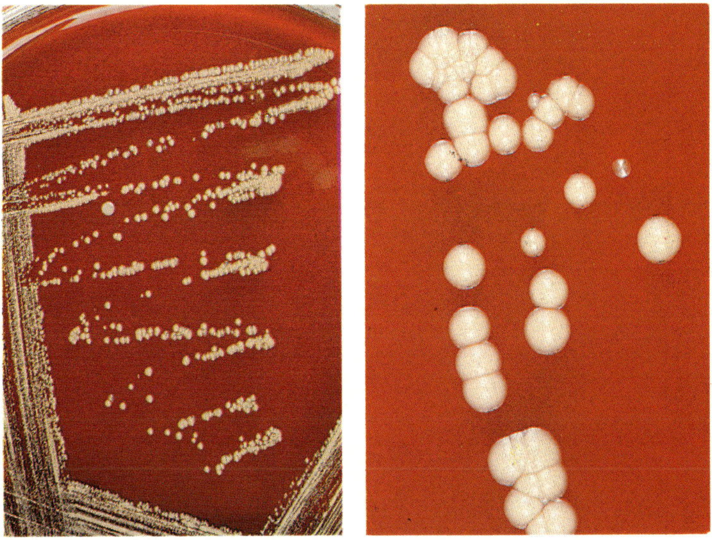

B

A, ×1. **B**, ×6.

Figure 108 'B. pulvifaciens' CCEB 638. 24 hours; 37°C.

B. popilliae and *B. larvae* – insect pathogens

Reference to these species is more to give full coverage of *Bacillus* than for the chief purpose of the atlas as a guide in the clinical laboratory. Bacteria of these species are fastidious and require growth factor-enriched (particularly thiamine) media for isolation, growth and maintenance in the laboratory. Media for biochemical tests must also be suitably enriched. They are thus relatively difficult to study, particularly the four varieties of *B. popilliae* (Milner, 1981) – var. *popilliae*, var. *lentimorbus*, var. *melolonthae* and var. *rhopaea* – which are especially fastidious. *B. larvae* can, in fact, be grown on general nutrient media containing added yeast extract (de Barjac, 1981). It is also difficult to induce sporulation by *B. popilliae* in laboratory media (Gordon *et al*, 1973).

These laboratory problems have made it difficult to acquire a great deal of knowledge about these species. They are established causative agents of certain insect diseases: *B. popilliae* infects the Scarabaeidae (beetles) and causes milky disease in the Japanese beetle (*Popillia japonica*) and other ground beetles – the larvae turn milky-white from the heavy production of *Bacillus* spores in the haemolymph (de Barjac, 1981). *B. larvae* is pathogenic for honey-bees causing 'foulbrood' (Bailey, 1981).

It is unlikely that *B. larvae* and *B. popilliae* will prove of importance in human or animal disease. In extensive tests to verify the safety of pest control preparations incorporating *B. popilliae* as the active ingredient, the species caused no ill effects in chickens, starlings, rats and monkeys when administered by various routes (Burges, 1981 and 1982).

Neither *B. popilliae* nor *B. larvae* grow on horse blood agar. It is possible that *B. larvae* might grow on or in certain yeast-extract containing media, but it is unlikely that either species would be found under the conventional conditions of a clinical laboratory. *B. popilliae*, it appears, must become laboratory-adapted before it will grow readily on even the most appropriate 'J' medium (Gordon *et al*, 1973).

B. larvae is closely related to 'B. pulvifaciens'; the two are differentiated by catalase reaction (*B. larvae*, negative; 'B. pulvifaciens', delayed weak-positive) and the inability of *B. larvae* to survive serial transfer in unsupplemented nutrient broth.

B. popilliae has been successfully put to commercial use in controlling the Japanese beetle, but *B. larvae*, being a pathogen of useful insects, does not lend itself to this type of use (de Barjac, 1981).

Our own experience was confined to three strains of *B. popilliae*, one of which (NRRL 2309) is shown in **Figures 109** and **110**. The information given above derives principally from the references cited.

109

A 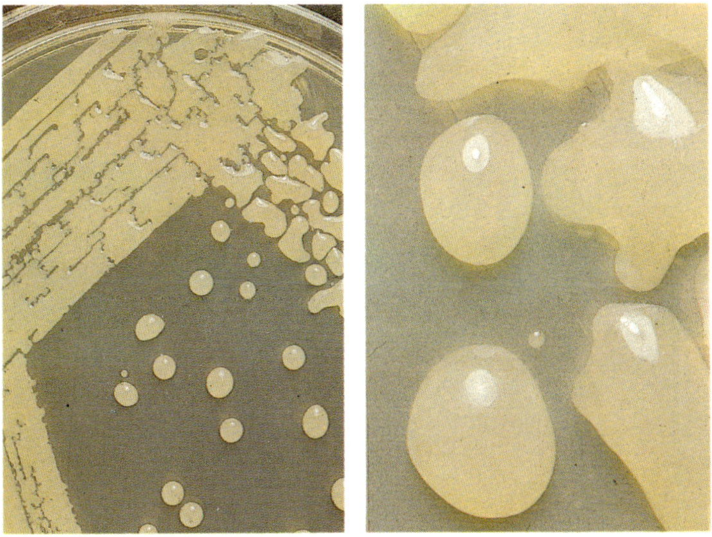 B

A, ×1. B, ×6.

Figure 109 ***B. popilliae*** **NRRL 2309.** 48 hours; 22°C.

This figure is included as representative of the catalase negative insect pathogens. The strain was grown in diphasic 'J' broth (2 days, 22°C on a water-bath shaker set at 50 cycles/min) and subcultured on to 'J' agar plates.

110

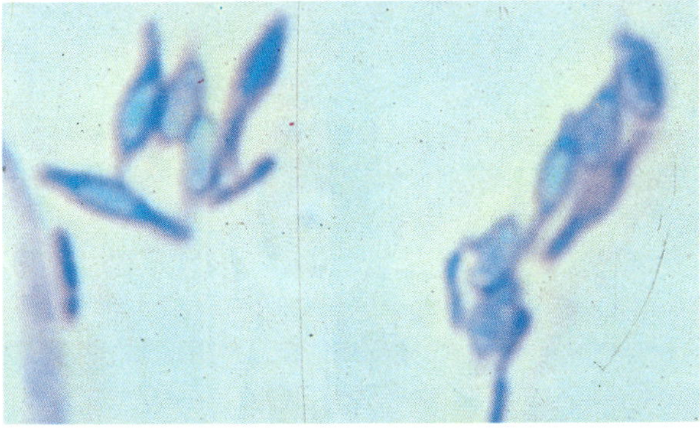

Figure 110 Vegetative cells and spores of *B. popilliae*. Phase contrast, ×1800.

The spores shown may have been either residual from the inoculum or, more likely, the 'weak' *in vitro* spores which form in 'J' medium. Absence of a parasporal body indicates they are incomplete (Dr B.N. Dancer, personal communication).

Morphological Group 3

Defining characteristics

Spores spherical, subterminal or terminal.
Bacillary body (sporangium) swollen.

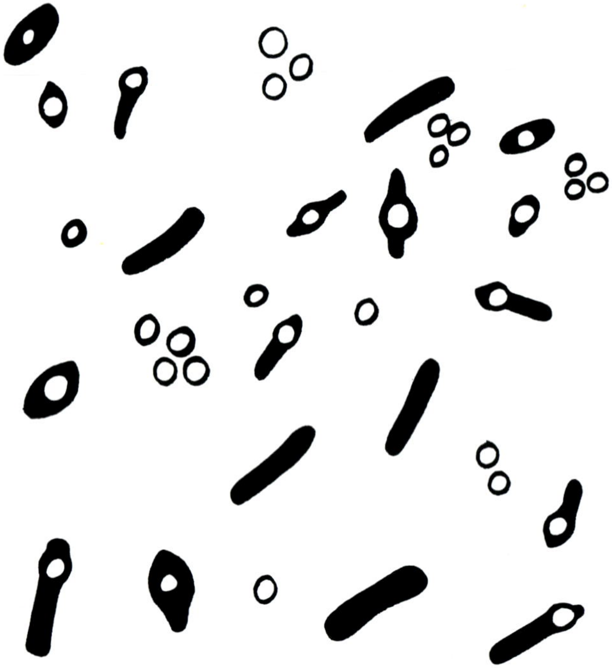

Group 3 species

1. *B. sphaericus*
2. *B. pasteurii**

*Strains maintained and examined with the addition of 1% urea to the media.

111

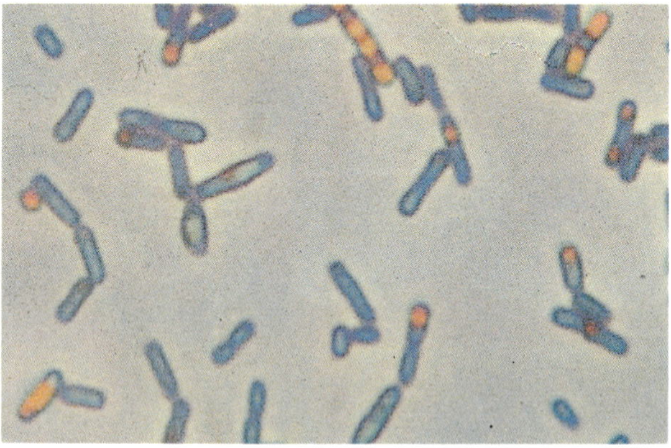

Figure 111 Vegetative cells and spores of *B. sphaericus*. Phase contrast, ×1600.

Bacillus sphaericus

The name of this species derives from its characteristically spherical spores; these are quite clear in the majority of cells in a smear such as seen in **Figure 111**.

Single rods (note that many of the cells in **Figure 111** have not completely divided) are of the order of $0.8 \times 3.5\,\mu m$. The spores swell the sporangia slightly (as in **Figure 111**) to extensively and are generally terminal or subterminal.

The majority of strains examined produced colonies 2–4 mm in diameter (24 hours; 37°C) which were low convex to convex with a circular, slightly irregular edge. These colonies were creamy-grey and had a butyrous consistency and smooth opaque surfaces. The Type strain, NCTC 10338/ATCC 14577 (**Figure 112**) produced slightly spreading colonies.

Haemolysis was apparent to one extent or another with all strains examined.

112

A B

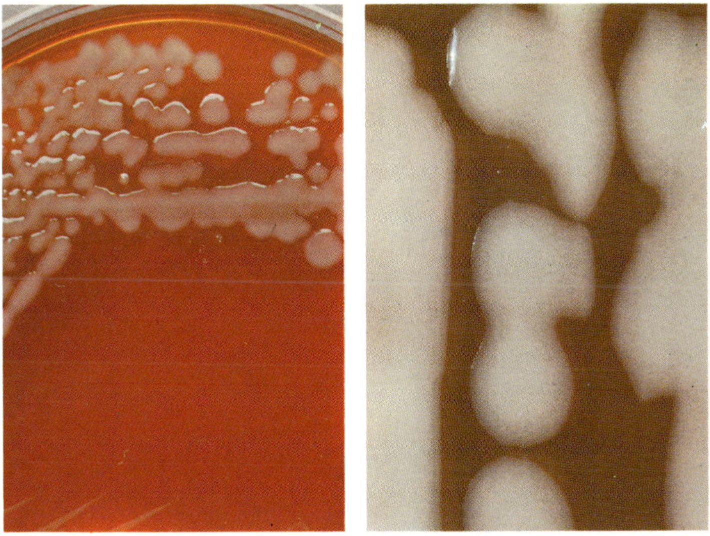

A, ×1. **B**, ×6.

Figure 112 ***B. sphaericus*** **NCTC 10338/ATCC 14577 (Type strain).** 24 hours; 37°C.

113

A B

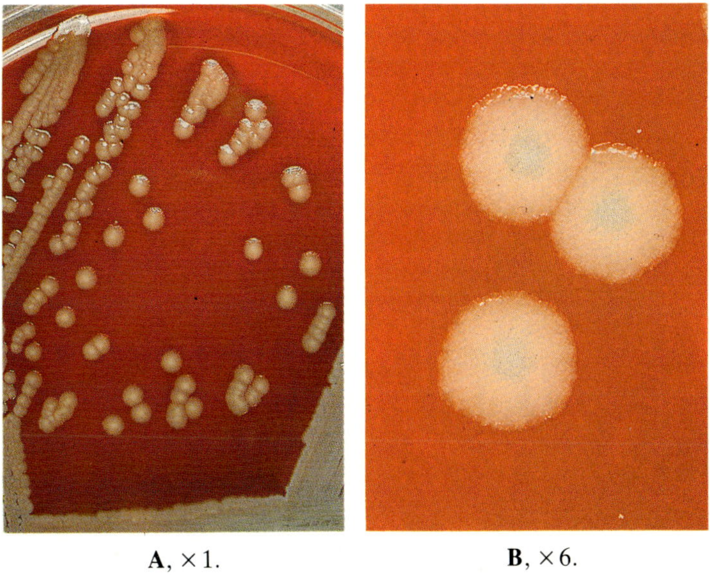

A, ×1. **B**, ×6.

Figure 113 ***B. sphaericus*** **NCTC 2609/ATCC 4525.** 24 hours; 37°C.

114

A

B

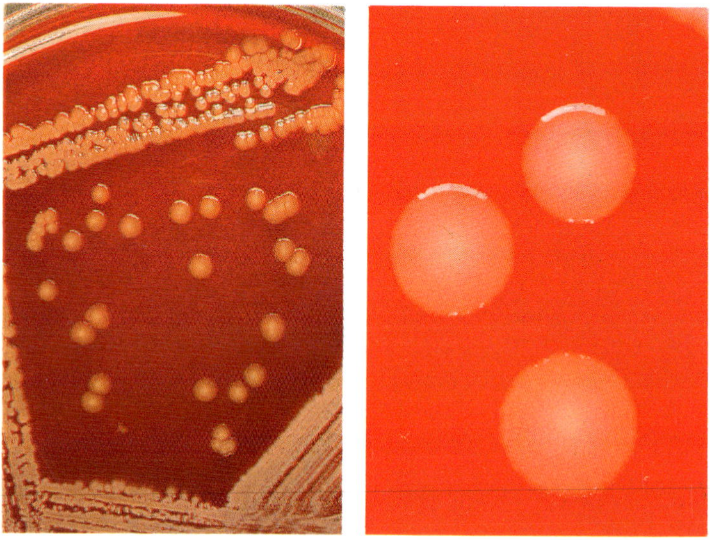

A, ×1. **B**, ×6.

Figure 114 ***B. sphaericus*** **NCTC 7582.** 24 hours; 37°C.

B. sphaericus is important as a pathogen of mosquito larvae. It is easily grown in laboratory culture media and, therefore, like *B. thuringiensis* but in contrast to *B. larvae*, and more particularly, *B. popilliae*, the other *Bacillus* insect pathogens, it has been relatively easy to study.

DNA homology studies have shown the species to fall into 5 differentiable groups, and all isolates entomotoxic for mosquito larvae have so far fallen into just one of these groups. The development of phage and flagellar (H) antigen typing systems has revealed that it is possible to differentiate the species further (de Barjac, 1981).

115

A 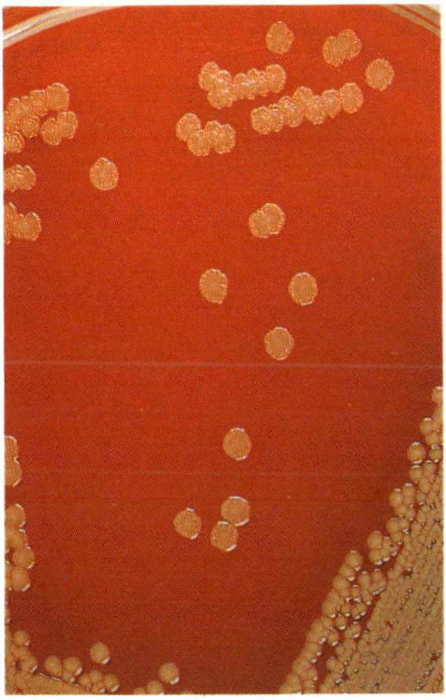 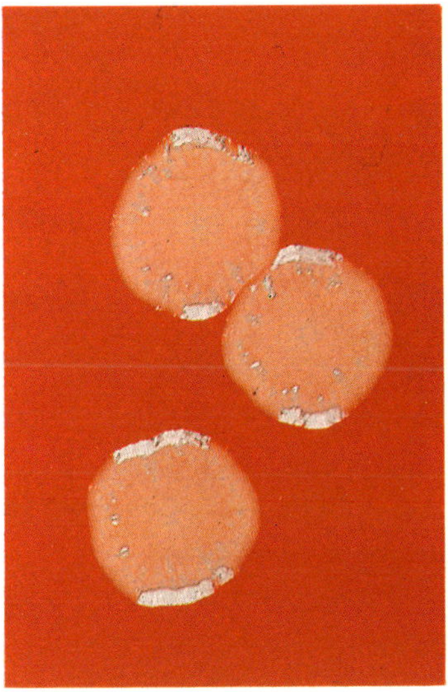 B

A, ×1. **B**, ×6.

Figure 115 ***B. sphaericus* NCTC 9602.** 24 hours; 37°C.

B. sphaericus has been implicated in human disease – meningitis, bacteraemia and endocarditis in a chronic alcoholic (Farrar, 1963), fatal meningitis with generalized Shwartzman reaction in a previously healthy adult (Allen and Wilkinson, 1969), fatal pulmonary pseudotumour in an adult with chronic asthma (Isaacson *et al*, 1976) and a small outbreak of food poisoning (Elter, 1966).

The species is not uncommon and like many of the *Bacillus* species, is found in soil, marine and fresh water sediments; it has been isolated from milk and other foods. It was also among the various *Bacillus* species found in the proprietary liquid antacids examined by Willemse–Collinet *et al* (1981).

One or two strains are of commercial value in production of useful enzymes or other biochemical products as well as those which are of developing interest in the realms of mosquito control.

116

A

B

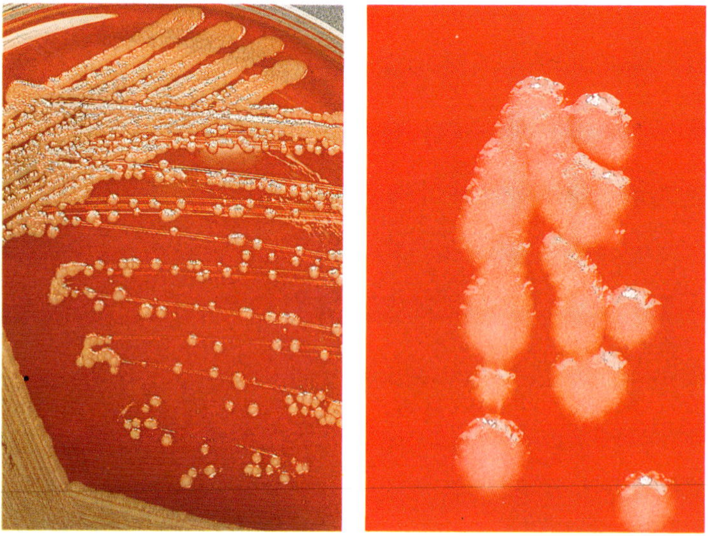

A, ×1. **B**, ×6.

Figure 116 ***B. sphaericus*** **NCTC 4825/ATCC 19394 (formerly *B. loehnisii*).** 24 hours; 37°C.

The mosquito strain 1593 and isolates available from the incidents referred to above have been tested extensively in a range of experimental animals. No illness was produced by inoculation of large numbers orally or parenterally, and it has been concluded that the isolates from the human infections may have been opportunistic (Burges, 1981 and 1982).

Strains NCTC 5896 and NCTC 7585, both originally designated *B. fusiformis*, differed in their colonial appearance from other strains examined. NCTC 5896 (**Figure 117**) produced large creamy-grey colonies, 5–8 mm in diameter (24 hours; 37°C), flat with an entire edge and almost transparent surface. NCTC 7585 again produced large colonies (**Figure 118**) 5–6 mm in diameter, which were grey-green, raised and had an irregular edge. These could have been mistaken for *B. cereus*.

Both strains were haemolytic.

Biochemically and otherwise, these strains were correctly placed within the species *B. sphaericus*.

117

A

B

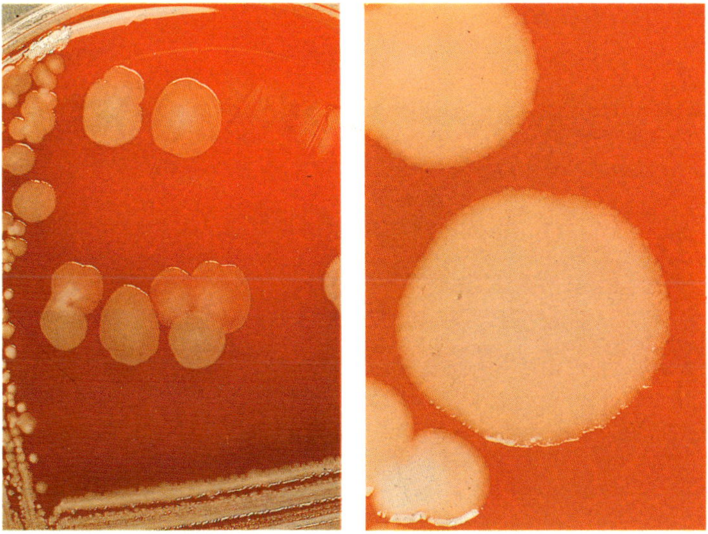

A, ×1. **B**, ×6.

Figure 117 ***B. sphaericus*** **NCTC 5896.** 24 hours; 37°C.

118

A

B

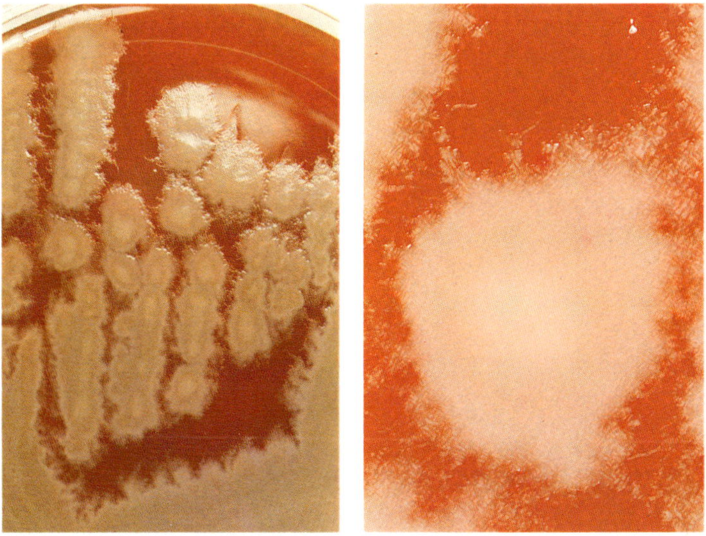

A, ×1. **B**, ×6.

Figure 118 ***B. sphaericus*** **NCTC 7585.** 24 hours; 37°C.

Bacillus pasteurii

This species is unlikely to be isolated on routine laboratory media as it requires urea or ammonium salt supplement in such media in order to grow. The maximum growth temperature may also be less than 37°C depending on the strain.

B. pasteurii is strongly 'ureaclastic' – that is, it converts urea to ammonium carbonate – and apparently does so faster than any other known bacterium (Gibson and Gordon, 1974). Either 1–2% urea or 1% ammonium salt (in the case of the latter, the optimal pH is 9) may be used as a supplement.

Colonies of the Type strain (**Figure 119**) on blood agar with 1% incorporated urea (48 hours; 22°C) measured 1–3mm in diameter, were convex and circular with an entire edge. The colonies were glossy, creamy-buff and markedly haemolytic.

B. pasteurii is quite closely related to *B. sphaericus*; its cells are similar in size to those of *B. sphaericus* and its spores are similarly round, terminal or subterminal, swelling the sporangium.

B. pasteurii has been isolated from soil, water and sewage.

119

A

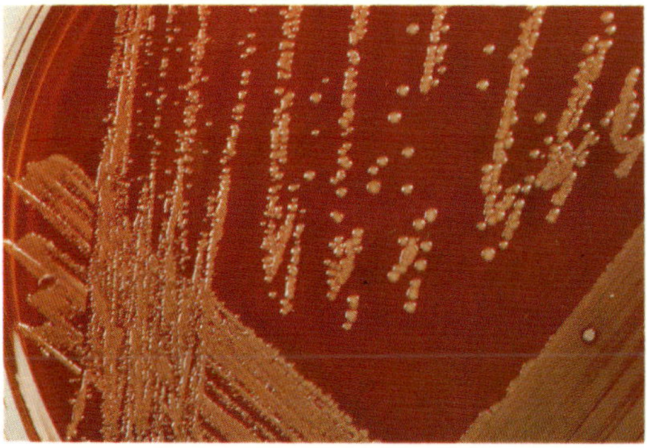

A, × 1. 48 hours; 22°C.

B

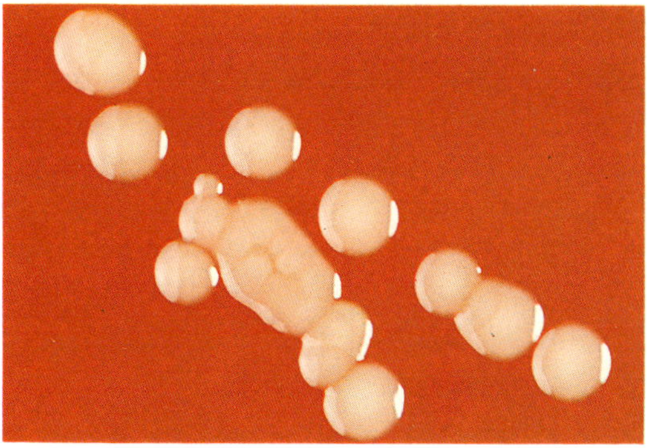

B, × 6. 48 hours; 22°C.

Figure 119 ***B. pasteurii*** **NCTC 4822/ATCC 11859 (Type strain).**

Unassigned strains

As was covered in the Introduction, after Smith *et al* (1952) and subsequently Gordon *et al* (1973) had divided the adequately represented strains of *Bacillus* into the species of Morphological Groups 1, 2 and 3, there remained a number of strains of aerobic spore-formers which did not obviously belong to the species within these groups. They felt unable to assign these to further definitive groups because the number of representative strains available in each species was too few for them to confidently prepare species descriptions. Gordon *et al* referred to these as 'unassigned strains' and 'poorly represented taxa'.

The majority of these species, i.e. those not in italics, were not included in the Approved Lists of Bacterial Names (1980), which essentially means that they await formal re-naming, and to date have no nomenclatural value. This, in turn, awaits the isolation of more representative strains within each species. On the basis that one of the most likely persons to encounter such further strains is the alert clinical or veterinary microbiologist, we are including in the following pages representatives of these 'unassigned strains'. It is hoped that this will lead to identification of further representative strains which, in turn, will enable clearer classification of this miscellaneous group of *Bacillus* species.

The unassigned strains were divided by Gordon *et al* (1973) into categories 1 to 5 on the basis of the apparent differences in their properties. We have designated them A to E to avoid confusion with Morphological Groups 1, 2 and 3 and have further subdivided group E on the basis of spore shape and ability to grow anaerobically.

All subgroups A to D strains grew in our 37°C hot room; those that are illustrated following incubation at 30°C merely grew better at this temperature than at 37°C.

Subgroup A

Defining characteristics

Spores oval.
Anaerobic growth positive.
Starch hydrolysis positive.
Fail to grow at 3°C.

Subgroup A species

1. 'B. apiarus'
2. 'B. filicolonicus'
3. 'B. thiaminolyticus'
4. *B. alcalophilus*

'Bacillus apiarus'

This species was originally isolated from honeybee larvae – hence its name – but it is not considered to be an insect pathogen. The species is not listed in the Approved Lists of Bacterial Names (1980), but two strains with this name are available from the ATCC. One of these is shown in **Figure 120**.

This strain (ATCC 29575) grew readily on blood agar. The colonies were uneven in shape and size; they had an irregular edge with an effuse elevation and smooth, tending to transparent surfaces. They were creamy-buff coloured and slightly haemolytic.

The spores were oval, central and swelled the sporangia; one is seen in the centre of **Figure 121**.

'B. apiarus' differs from the members of Morphological Group 2 (those whose sporangia are swollen by ellipsoidal spores) as a whole in that it is urease positive.

120

A

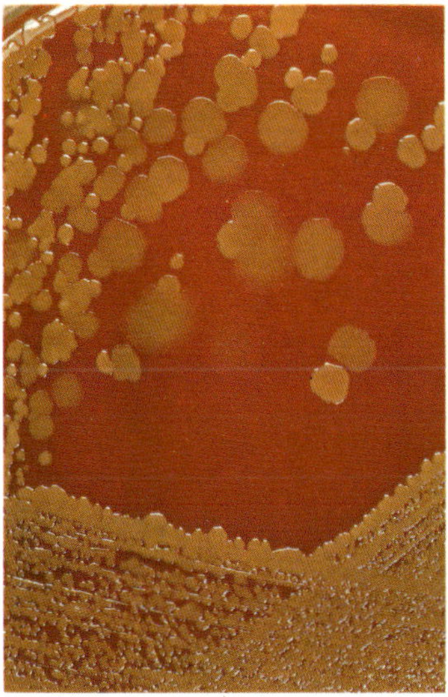

B

A, ×1. **B**, ×6.

Figure 120 'B. apiarus' ATCC 29575. 24 hours; 30°C.

121

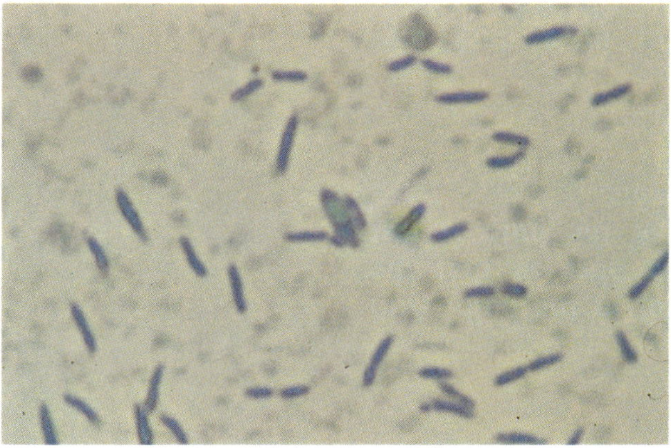

Figure 121 Vegetative cells with one sporulating cell of 'B. apiarus'. Phase contrast. ×1600.

‘Bacillus filicolonicus’

Strain ATCC 14413 (**Figure 122**) is the original isolate of this species and was isolated from sea water. Apparently, strains of ‘B. filicolonicus’ are not uncommon in marine habitats (Priest *et al*, 1981). Nevertheless, the species is not listed in the Approved Lists of Bacterial Names (1980).

The name means literally ‘thread-like colonies’ (*filum* = Latin, thread) and as seen in **Figure 122B**, the colonies have an irregular, almost rhizoid edge from which, presumably they derived this name. The colonies were raised and were creamy-grey with a shiny, opaque surface.

Slight haemolysis was apparent on the blood agar.

The cells were fairly long slender rods and the spores were oval swelling the sporangium in a subterminal or terminal position (**Figure 123**).

‘B. filicolonicus’ is most readily differentiated from Morphological Group 2 species (those with sporangia swollen by ellipsoidal spores) by its high salt tolerance. It is capable of growth in 10% NaCl broth.

122

A B

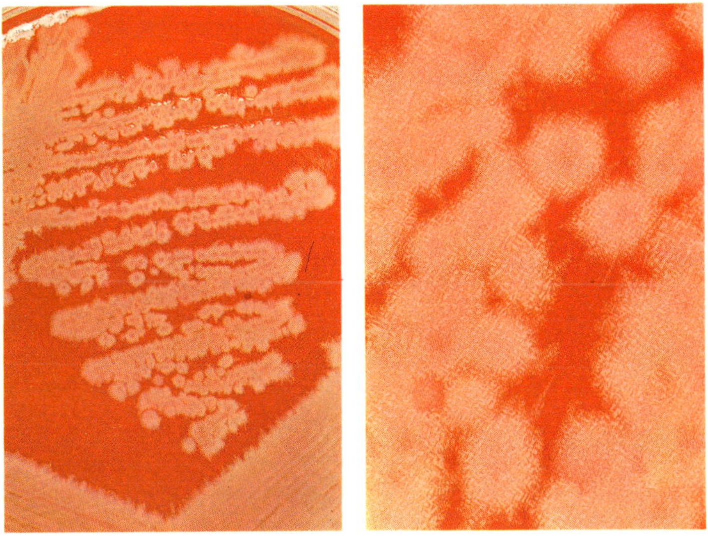

A, ×1. B, ×6.

Figure 122 'B. filicolonicus' ATCC 14413. 24 hours; 30°C.

123

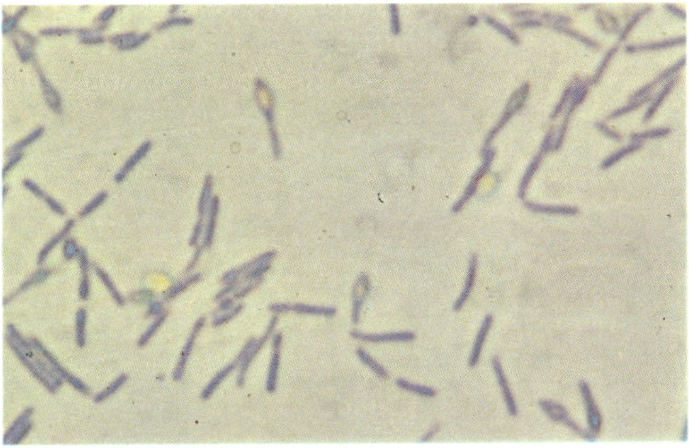

Figure 123 Vegetative cells and spores of 'B. filicolonicus'. Phase contrast, ×1200.

124

A

B

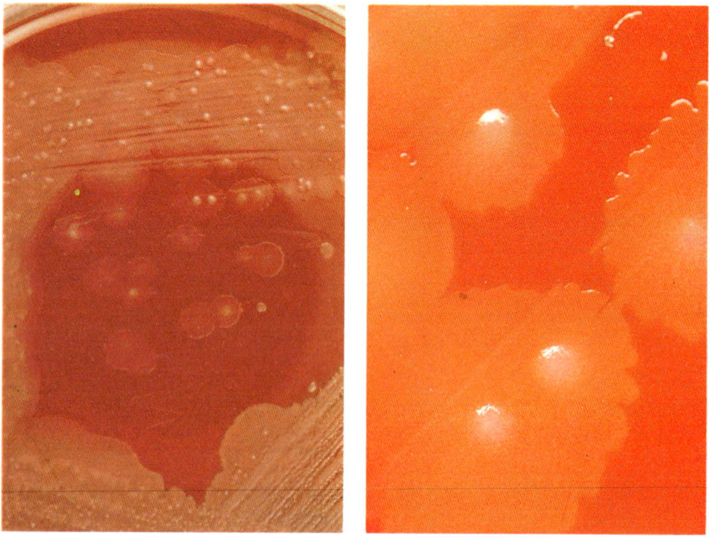

A, ×1. **B**, ×6.

Figure 124 'B. thiaminolyticus' NCTC 10760. 24 hours; 37°C.

'Bacillus thiaminolyticus'

This species was so-named because the earliest species to be isolated decomposed thiamine actively although the value of thiaminase activity as a criterion of speciation later fell into doubt (Gordon *et al*, 1973).

It appears to be most closely related to *B. alvei* but insufficient strains are available to allow this to be definitively established. The biochemical differences from *B. alvei* make it necessary to leave this species among the unassigned group for the time being.

The colonies (24 hours; 37°C) were large (4–6mm), spreading and translucent with an irregular edge and a raised convex centre – 'fried egg' appearance.

Slight haemolysis was produced on the blood agar.

The spores were oval and swelled the sporangium in the subterminal or terminal position.

Gibson and Gordon (1974) give the source of this organism as being human faeces.

125

A 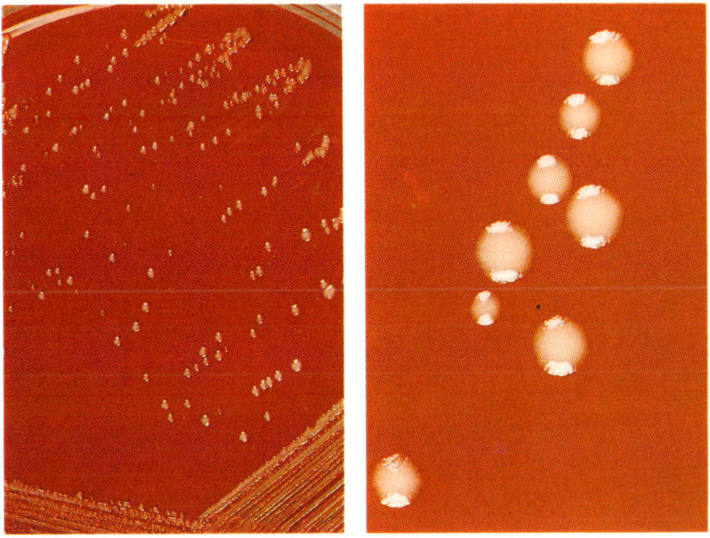 B

A, ×1. **B**, ×6.

Figure 125 ***B. alcalophilus*** **NCTC 4553/ATCC 27647 (Type strain).** 24 hours; 37°C.

Bacillus alcalophilus

The name means literally 'loving alkaline' and the species was originally characterized by a marked tolerance to alkali and, according to Gibson and Gordon (1974), an inability to grow on media at pH 7.

Strain NCTC 4553 (**Figure 125**) did, however, grow quite readily on blood agar and produced circular, entire, convex colonies, 1–3 mm in diameter (24 hours; 37°C), creamy white in colour and with a moist consistency.

There was a slight haemolysis on the blood agar.

The rod size was of the order of $0.7 \times 4\,\mu m$ and spores were oval, subterminal, swelling the sporangia slightly.

As well as its alkalophilic properties (ability to tolerate pHs of up to 11), the species is unlike Morphological Group 2 species (those whose sporangia are swollen by ellipsoidal spores) in being unable to ferment glucose ammonium salt sugar – at least to the extent of producing a visible acid reaction.

Strain NCTC 4553 was isolated from human faeces.

Subgroup B

Defining characteristics

Spores oval.
Anaerobic growth negative.
Starch hydrolysis positive.
Fail to grow at 3°C.

Subgroup B species

5. 'B. cirroflagellosus'
6. 'B. chitinosporus'
7. *B. lentus*

126

A B

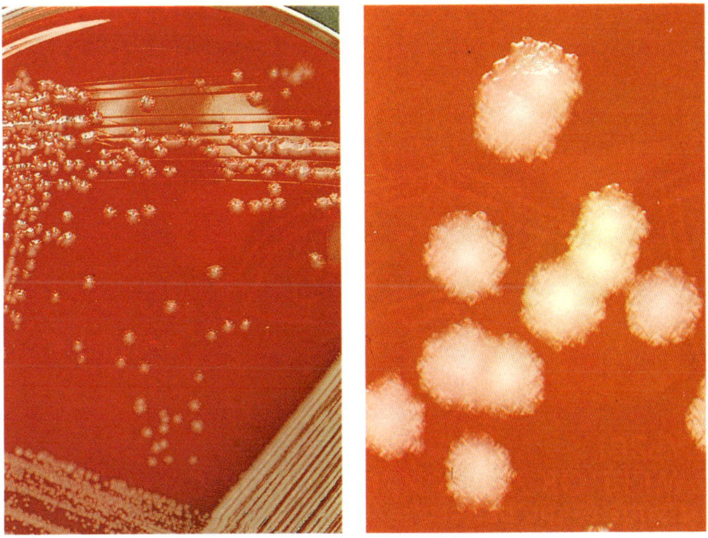

A, ×1. B, ×6.

Figure 126 **'B. cirroflagellosus' ATCC 14411.** 24 hours; 30°C.

'Bacillus cirroflagellosus'

Strain ATCC 14411 (**Figure 126**) is the original strain which was isolated from marine mud. But despite this, it does not have a high salt tolerance and will not grow in 5% NaCl broth.

The strain produced low convex circular colonies with an irregular edge and measured 2–3 mm in diameter (24 hours; 30°C). The colonies were greyish-white and produced slight haemolysis after prolonged incubation.

The cells were quite small – in the order of $0.6 \times 3\,\mu m$; the spores were oval, central and swelled the sporangia.

This species has neither a close resemblance to any other species nor any specially distinctive property (Gibson and Gordon, 1974). According to these authors, 30°C is its maximum growth temperature, so it is unlikely to be encountered in the clinical laboratory with the customary 37°C hot incubation conditions. However, we did observe growth on blood agar plates placed for 24 hours in our 37°C hot room with both strain ATCC 14411 and an isolate of our own tentatively identified as 'B. cirroflagellosus'.

127

A

B

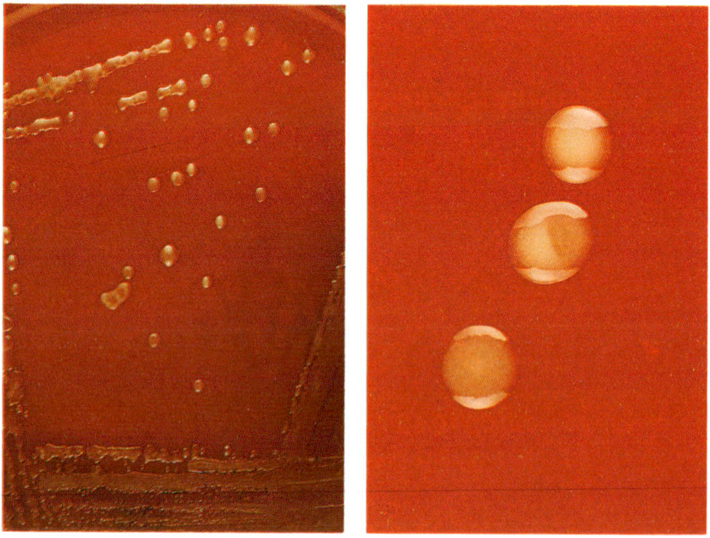

A, ×1. B, ×6.

Figure 127 'B. chitinosporus' NCIB 9652/ATCC 19986. 24 hours; 30°C.

'Bacillus chitinosporus'

This species is not listed in *Bergey's Manual* (Gibson and Gordon, 1974) since Gordon *et al* (1973) felt unable to evaluate the authenticity of ATCC 19986 (**Figure 127**) which bears the name.

It is, in fact, not clear whether other strains have come to light. The properties listed in **Table 4** are those of this one strain. Colonies were 2–3mm in diameter, round, convex, entire with a moist, almost mucoidal consistency. Haemolysis was apparent in older cultures.

As seen in **Figure 128**, the strain produced unusual chains of undivided cells in which the developing spores could be seen swelling the sporangia. The diameter of the fully developed spore greatly exceeded that of the vegetative cell.

This strain was isolated from soil.

128

A

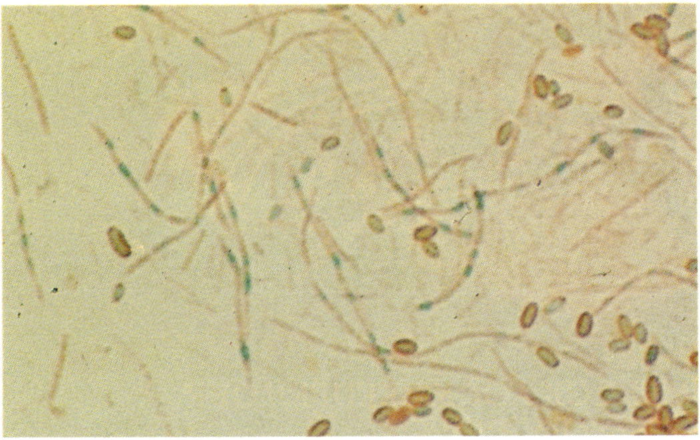

A, Malachite green stain. × 1200.

B

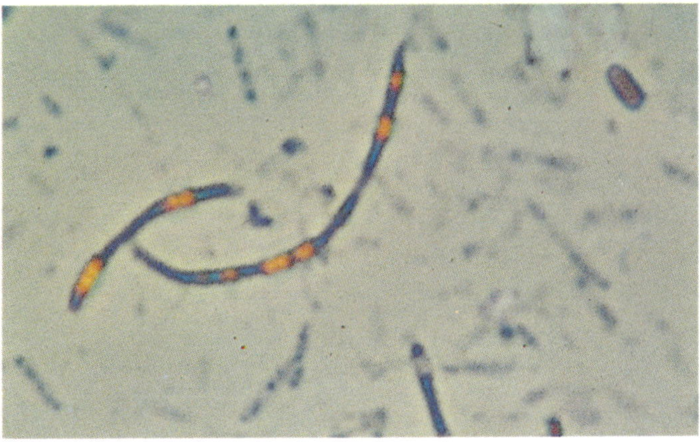

B, Phase contrast. × 3000.

Figure 128 Vegetative cells and spores of 'B. chitinosporus'.

Bacillus lentus

This species is closely related to *B. firmus* but, on the basis of DNA homology studies, Priest (1981) felt that *B. lentus* was a valid species in its own right and it is listed in the Approved Lists of Bacterial Names (1980).

For practical purposes, the two species are distinguished by nitrate reduction, decomposition of casein and liquefaction of gelatin (**Tables 2** and **4**). However, intermediates between the two species exist.

The two strains illustrated in **Figures 129** and **130** were quite different in colonial appearance although they were identical biochemically.

Strain NCTC 4824 (**Figure 129**) produced low convex, circular colonies with an irregular 'fuzzy' edge measuring 3–4mm in diameter (24 hours; 30°C). The colonies were cream-buff with a butyrous consistency. Slight haemolysis was apparent upon incubation for a further 24 hours.

Strain CCEB 630 produced circular, entire, convex colonies, 1–3mm in diameter (24 hours; 30°C). These colonies were creamy-white, shiny and butyrous. Slight haemolysis was apparent at 48 hours.

The cells were small (approximately $0.75 \times 2.5\,\mu m$); the spores were oval, central and, true to their similarity to Morphological Group 1 *B. firmus*, did not swell the sporangia.

The described strains of *B. lentus* have been isolated from soil. The species was also listed by Seenappa and Kempton (1981) as one of the predominant *Bacillus* isolates in samples they had analysed of black pepper, red pepper and ginger.

129

A B

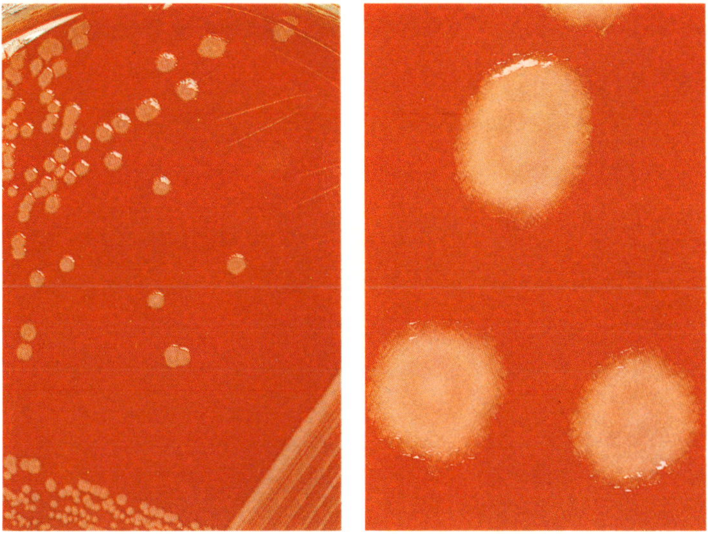

A, ×1. **B**, ×6.

Figure 129 ***B. lentus*** **NCTC 4824/ATCC 10840 (Type strain).** 24 hours; 30°C.

130

A B

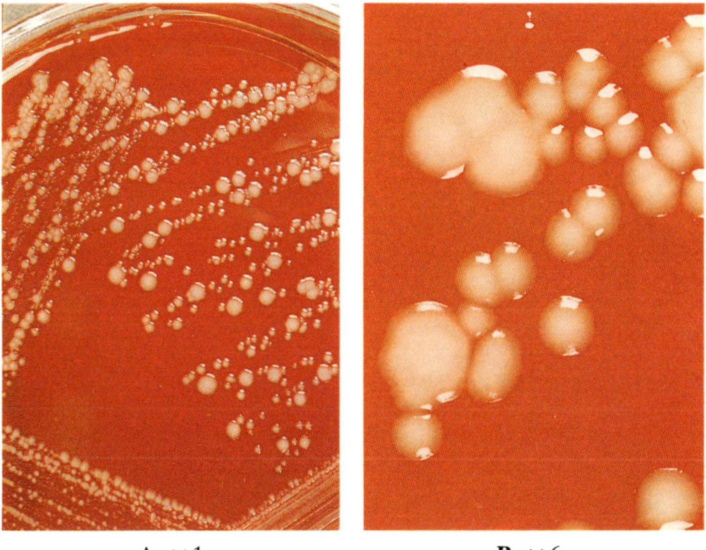

A, ×1. **B**, ×6.

Figure 130 ***B. lentus*** **CCEB 630/ATCC 10841.** 24 hours; 30°C.

Subgroup C

Defining characteristics

Spores oval.
Anaerobic growth negative.
Starch hydrolysis negative.
Fail to grow at 3°C.

Subgroup C species

8. *B. badius*
9. 'B. aneurinolyticus'
10. 'B. macroides'
11. 'B. freundenreichii'

131

A 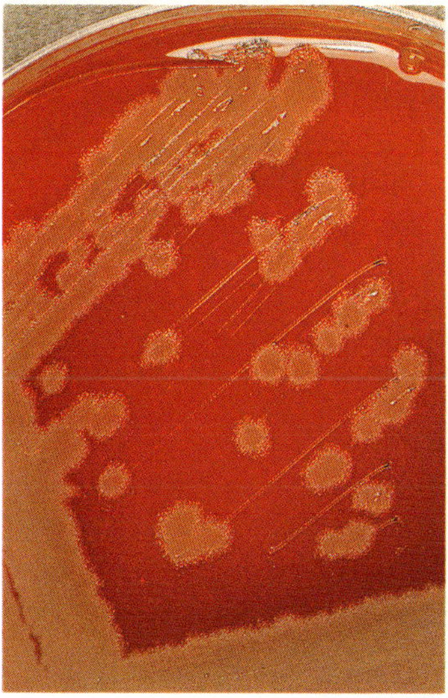 B 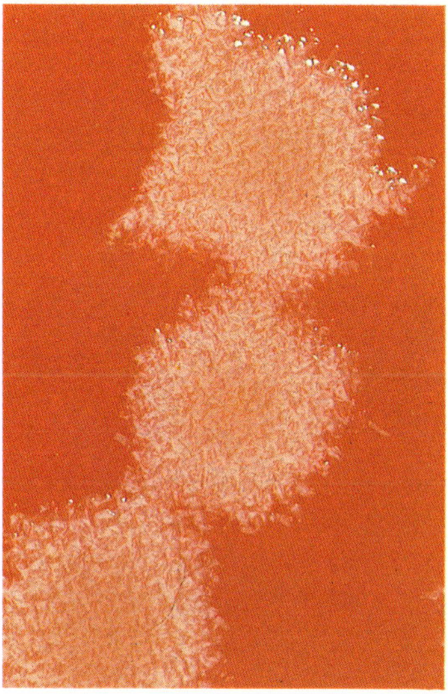

A, ×1. B, ×6.

Figure 131 ***B. badius*** **NCTC 10333/ATCC 14574 (Type strain).** 24 hours; 37°C.

Bacillus badius

The name has a Latin derivation and means 'chestnut brown', resulting presumably from the colour of the colonies on the medium used when it was first isolated in 1919.

Of the species of Morphological Groups 1, 2 and 3, it is most closely related to *B. brevis* (Morphological Group 2). However, one of the distinguishing characteristics between the two species is that the spores of *B. badius* do not swell the sporangium – a Morphological Group 1 characteristic. A second differential characteristic is the halophilic nature of *B. badius* (**Tables 3** and **4**). It differs from Morphological Group 1 species in general in being unable to ferment the four ammonium salt sugars listed in **Tables 2** and **4**.

Gibson and Gordon (1974) describe the colonies of the Type strain (**Figure 131**) as having a folded hair structure with rhizoid outgrowth.

B. badius has apparently been isolated infrequently from faeces, dust, foods and marine habitats. Willemse–Collinet *et al* (1981) isolated a strain from a sample of a proprietary liquid antacid.

'Bacillus aneurinolyticus'

Like *B. badius*, the closest relative of 'B. aneurolyticus' among the species of the three morphological groups is *B. brevis*; the only definite distinction, according to Gibson and Gordon (1974) is the lack of action of 'B. aneurinolyticus' on casein although some *B. brevis* strains are known to be weak or negative in this test.

'B. aneurinolyticus' is a biochemically inactive species with only nitrate reduction being positive (**Table 4**).

The species name means 'digesting thiamine' (aneurine = thiamine); decomposition of thiamine has not been reported with *B. brevis*. 'B. aneurinolyticus' is distinct from 'B. thiaminolyticus'.

Strain ATCC 12856 was distributed by the first persons to describe it as a representative strain. It was isolated from human faeces.

The strain produced round, low convex colonies, 4–5 mm in diameter (3 days; 37°C), with an irregular edge and raised central area giving a 'poached egg' appearance (**Figure 132**). Colonies were grey-white and produced a slight haemolysis on blood agar.

The vegetative cell size was approximately $0.9 \times 4\,\mu m$. The spores were oval, central or subterminal and markedly swelled the sporangia (**Figure 133**).

132

A

B

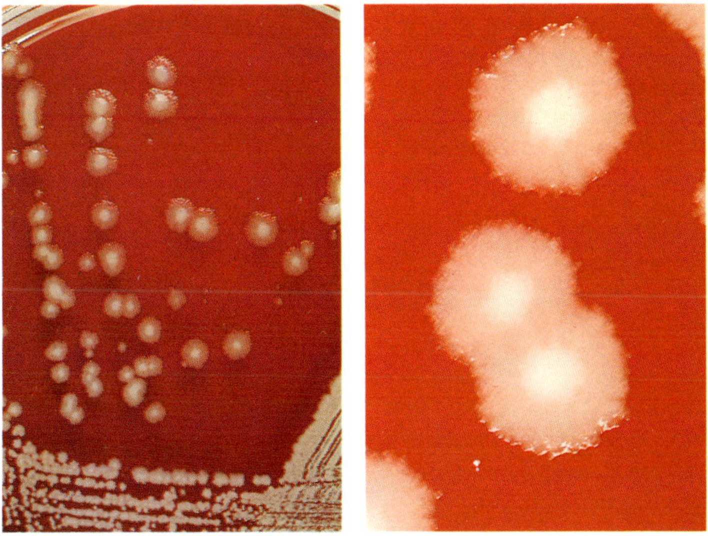

A, ×1. **B**, ×6.

Figure 132 'B. aneurinolyticus' IAM 1077/ATCC 12856. 3 days; 37°C.

133

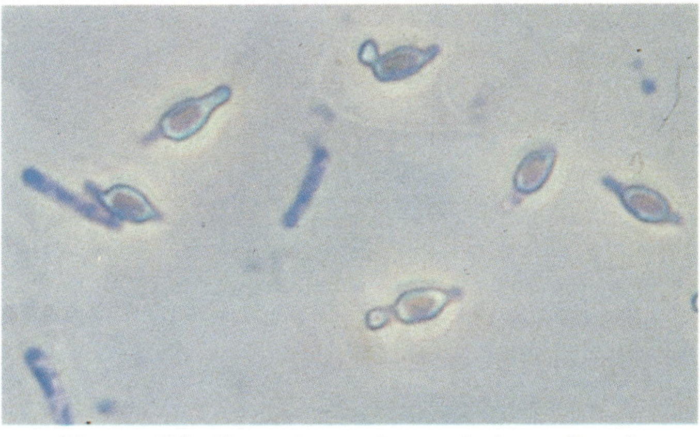

Figure 133 Vegetative and sporulating cells of 'B. aneurinolyticus'. Phase contrast. ×1800.

'Bacillus macroides'

In *Bergey's Manual* (Gibson and Gordon, 1974), it is suggested that the closest relative of 'B. macroides' among the species of Morphological Groups 1, 2 and 3 is *B. sphaericus*. However, the spores of 'B. macroides' are oval (**Figure 134**) and, according to Priest (1981), the genome base compositions indicate there may be reason to regard the two species as distinct.

Strain NCIB 8796 was proposed by the first workers to describe the species as the representative strain, but the species is not listed in the Approved Lists of Bacterial Names (1980).

This strain produced colonies 3–4 mm in diameter (24 hours; 37°C), round, umbonate and with a slightly irregular edge. The colonies were moist, butyrous and yellow-cream in colour. Slight haemolysis was apparent around the colonies.

The sources from which this species have been isolated are listed by Gibson and Gordon (1974) as cow dung and plant material decaying in water.

134

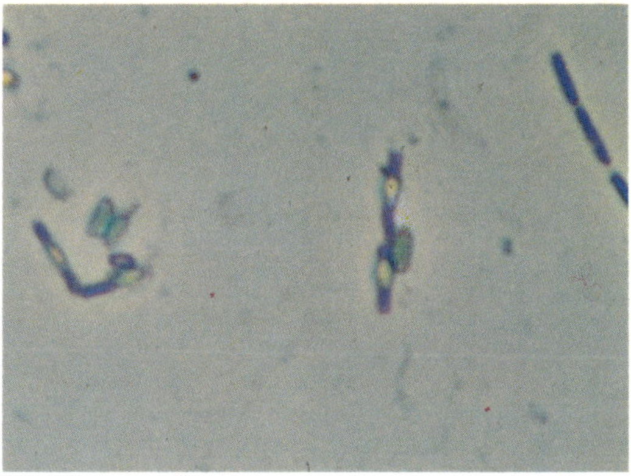

Figure 134 Vegetative and sporulating cells of 'B. macroides'. Phase contrast. ×1300.

135

A

B

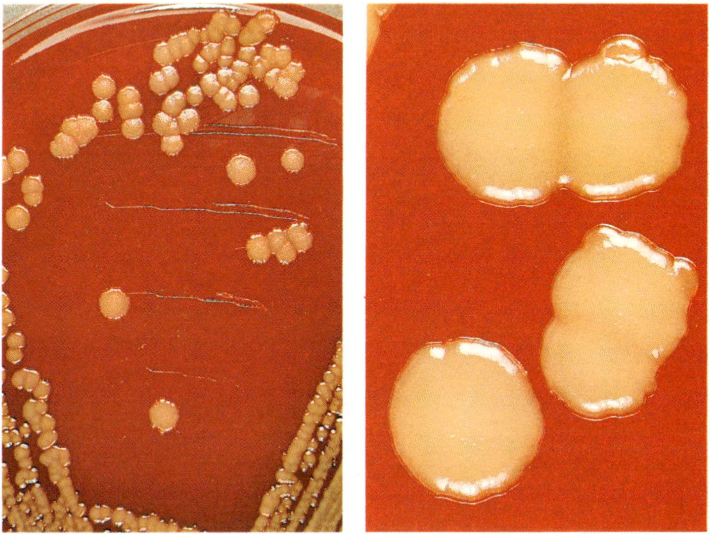

A, ×1. **B**, ×6.

Figure 135 'B. macroides' NCIB 8796/ATCC 12905. 24 hours; 37°C.

136

A

B

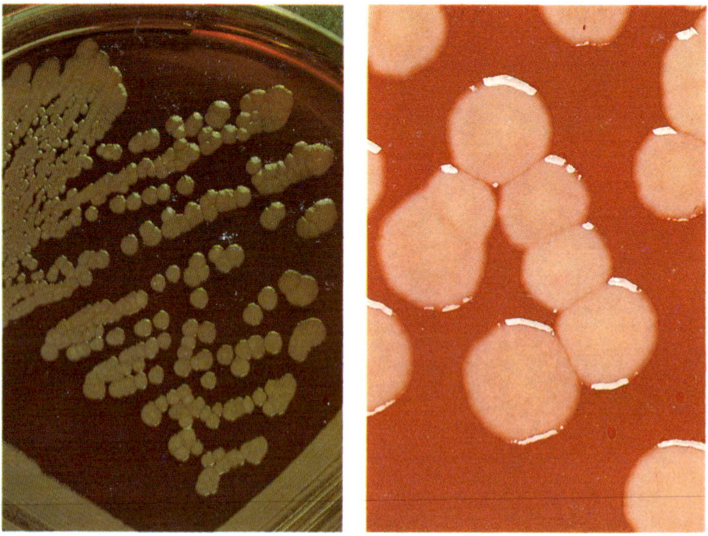

A, ×1. **B**, ×6.

Figure 136 'B. freundenreichii' NCTC 4823/ATCC 7053. 24 hours; 37°C.

'Bacillus freundenreichii'

According to *Bergey's Manual* (Gibson and Gordon, 1974), the nearest relative to 'B. freundenreichii' among the Morphological Groups 1, 2 and 3 species is *B. brevis* and it was suggested that it might be an intermediate between *B. brevis* and *B. pasteurii*.

Strain NCTC 4823 (**Figure 136**) was isolated from soil. It produced slightly raised, circular, entire colonies measuring 2–3mm in diameter. They were grey-green, moist and butyrous producing marked haemolysis on blood agar.

Subgroup D

Defining characteristics

Spores oval or spherical.
Anaerobic growth variable.
Starch hydrolysis positive.
Fail to grow at 3°C.

Subgroup D species

12. *B. pantothenticus*
13. 'B. epiphytus'

Bacillus pantothenticus

Generally, isolates of this species have been from soil, although it was found by Willemse–Collinet *et al* (1981) in 9 samples of proprietary liquid antacids.

The species is so-named because of its requirement for pantothenic acid as a growth factor. As reviewed by Logan and Berkeley (1981), it has been variously placed as being related to *B. coagulans* (Morphological Group 1) and *B. circulans* and *B. stearothermophilus* (Morphological Group 2). Since both oval and spherical spores were produced, Smith *et al* (1952) considered *B. pantothenticus* as being intermediate between Morphological Groups 2 and 3.

The organism is halophilic, growth being improved by addition of 4% NaCl to media for its growth and testing and it is able to grow well in 10% salt broths.

The strains examined produced circular, irregular, low-convex colonies 1.5–3mm in diameter (24 hours; 37°C). The colonies were whitish or creamy-buff and non-haemolytic at 24 hours. Slight haemolysis became apparent on prolonged incubation.

The vegetative cells were approximately $0.6 \times 2.3\,\mu m$. The spores we saw appeared to be spherical, usually subterminal and swelling the sporangia.

A B 137

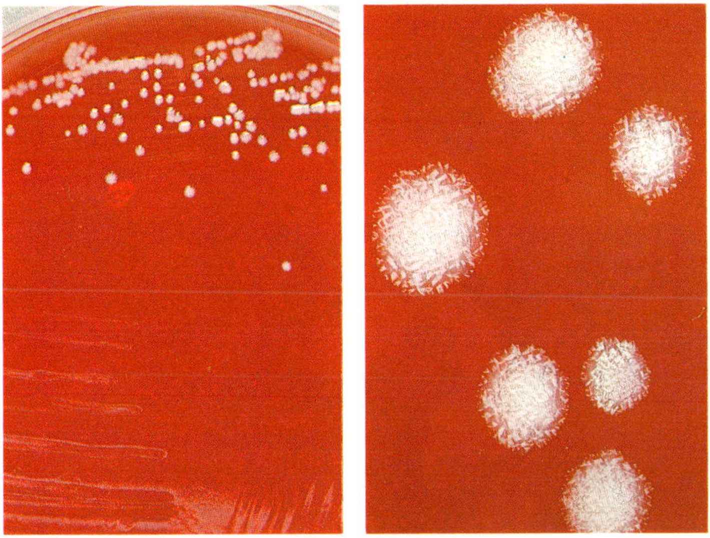

A, ×1. **B**, ×6.

Figure 137 ***B. pantothenticus*** NCTC **8162/ATCC 14567 (Type strain).** 24 hours; 37°C.

A B 138

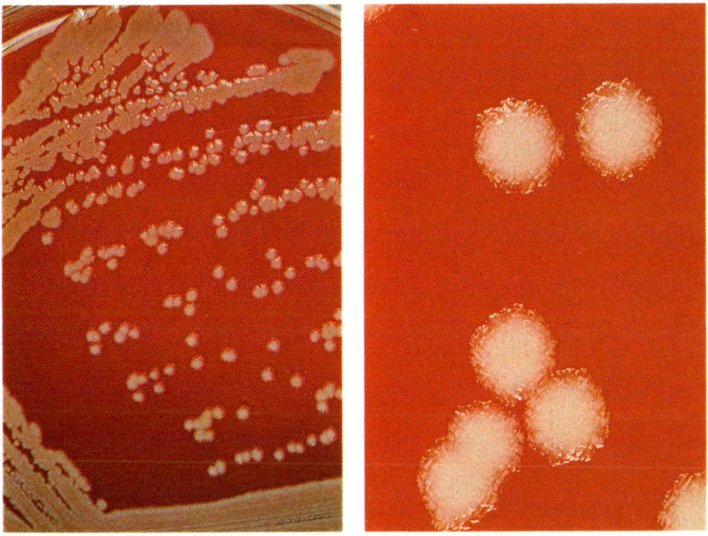

A, ×1. **B**, ×6.

Figure 138 ***B. pantothenticus*** **NCTC 8124.** 24 hours; 37°C.

‘Bacillus epiphytus’

The name of this species has a Greek derivation and means ‘growing on a plant’. ATCC 14412 (**Figure 139**) was the original strain and was isolated from marine phytoplankton.

It is halophilic and grows in 10% NaCl broths. It shares this halophilic property and production of a mixture of round and oval spores with the other subgroup D species, *B. pantothenticus*. The two are primarily distinguished by ability to grow anaerobically (*B. pantothenticus* = positive; ‘B. epiphytus’ = negative – **Table 5**).

The nearest relative among the main morphological groups appears to be *B. firmus*.

The colonies of strain ATCC 14412 (**Figure 139**) were circular, entire, convex, and measured 1–2mm in diameter (24 hours; 37°C). They were moist and creamy-yellow and gave rise to slight haemolysis.

The cell size was approximately $0.8 \times 2.5\,\mu m$. The spores were centrally or subterminally placed and occasionally swelled the sporangia. The mixture of round and oval spores can be seen in **Figure 140**.

139

A

B

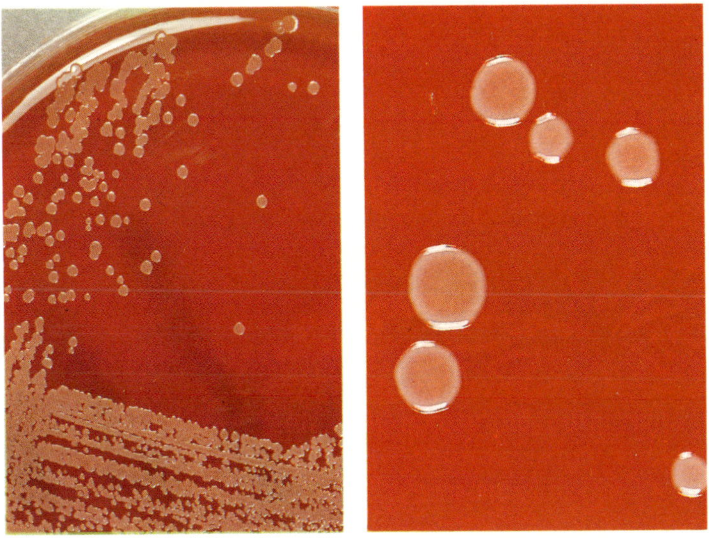

A, ×1. **B**, ×6.

Figure 139 'B. epiphytus' ATCC 14412. 24 hours; 37°C.

140

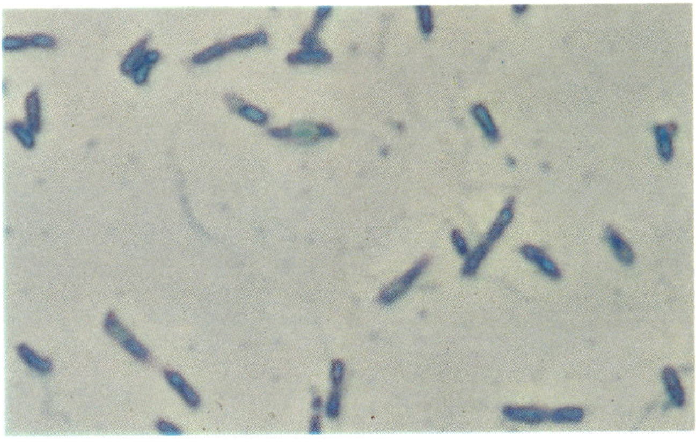

Figure 140 Vegetative and sporulating cells of 'B. epiphytus'. Phase contrast. ×1300.

Psychrophiles

Subgroup E of the unassigned strains comprises the psychrophilic *Bacillus* species. They are listed by Gordon *et al* (1973) as unassigned because, once again, of the lack of sufficient representative strains to permit definitive species descriptions. They do stand apart, however, from other *Bacillus* species; ability to grow at 3°C or below is an uncommon property within the genus as a whole.

We have subdivided the groups into Subgroups E_1 and E_2 on the basis of their spore shape and ability to grow anaerobically.

The maximum growth temperatures of 'B. aminovorans' and 'B. psychrosaccharolyticus' are in the region of 35°C; the others have maximum growth temperatures of 25–30°C. Thus, although they do grow on blood agar, it seems unlikely that they will be encountered in clinical contexts.

Subgroup E_1

Defining characteristics

Spores spherical.
Anaerobic growth negative.
Starch hydrolysis variable.
Grow at 3–5°C.

Subgroup E_1 species

14. 'B. aminovorans'
15. *B. globisporus*
16. *B. insolitus*
17. 'B. psychrophilus'

141

A

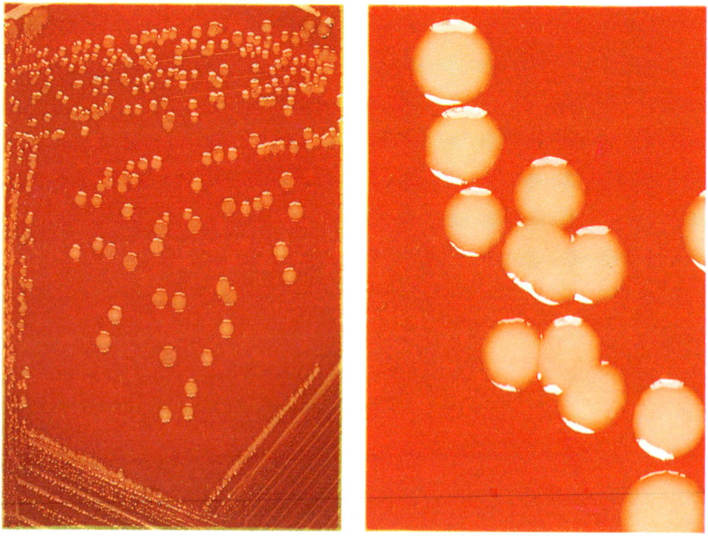

B

A, ×1. **B**, ×6.

Figure 141 'B. aminovorans' NCIB 8292/ATCC 7046. 48 hours; 22°C.

'Bacillus aminovorans'

At 22°C and after 48 hours, strain NCIB 8292 produced colonies measuring 1–3 mm in diameter and which were circular and convex with an entire edge. The colonies were moist, butyrous and creamy-yellow. Haemolysis was marked.

The cell size was of the order of $1.2 \times 2.5\,\mu m$, and the spores were round, central and did not swell the sporangia.

142

A

B

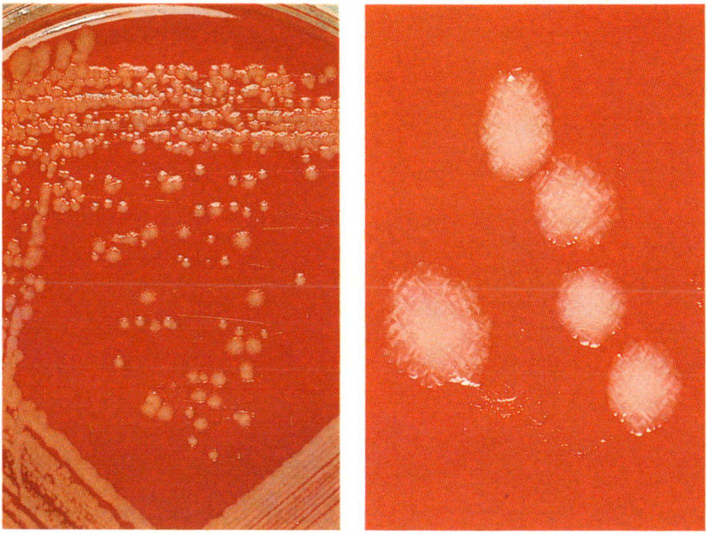

A, ×1. **B**, ×6.

Figure 142 ***B. globisporus*** **CCM 2119.** 48 hours; 22°C.

Bacillus globisporus

This species was named for its spherical spores (*globus* = Latin, sphere). Its nearest relative in the main morphological groups is *B. sphaericus*.

Colonies of *B. globisporus* strain CCM 2119 were circular, slightly raised with an irregular edge and measured 1–3 mm in diameter after 48 hours at 22°C. Slight haemolysis was apparent around the colonies at this time.

The cells measured $0.9 \times 3.2\,\mu m$ on average and contained round spores in a subterminal position; these swelled the sporangia.

Two subspecies are listed in the Approved Lists of Bacterial Names (1980): *B. globisporus* ss *globisporus* and *B. globisporus* ss *marinus*.

143

A

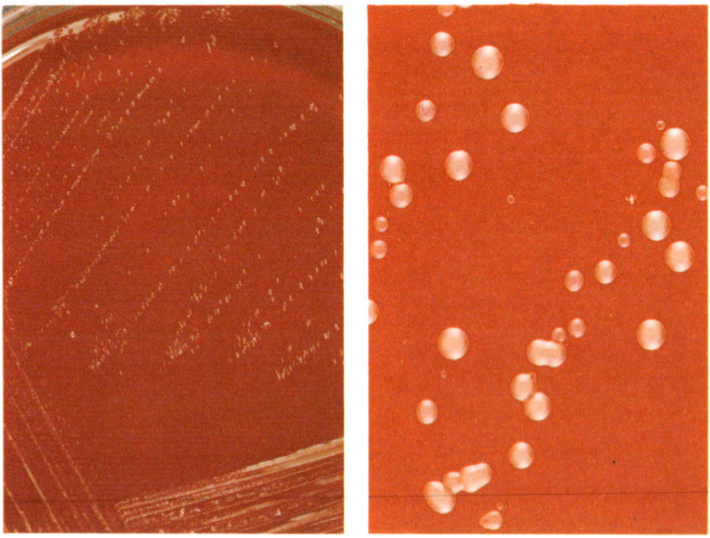

B

A, ×1. **B**, ×6.

Figure 143 ***B. insolitus*** **CCM 2175.** 48 hours; 22°C.

Bacillus insolitus

The colonies produced by *B. insolitus* strain CCM 2175 were minute after 48 hours' incubation at 22°C (**Figure 143**), so the morphology of the colonies after 5 days' incubation at 22°C is also shown (**Figure 144**).

The colonies were circular, entire, convex and had produced just slight haemolysis at 5 days. They were shiny, white, moist and had a butyrous consistency.

Biochemically, it is a notably inactive species.

The cells were approximately $1 \times 2.4\,\mu m$. The spores were round and lay in a terminal or subterminal position not swelling the sporangia (**Figure 145**).

A 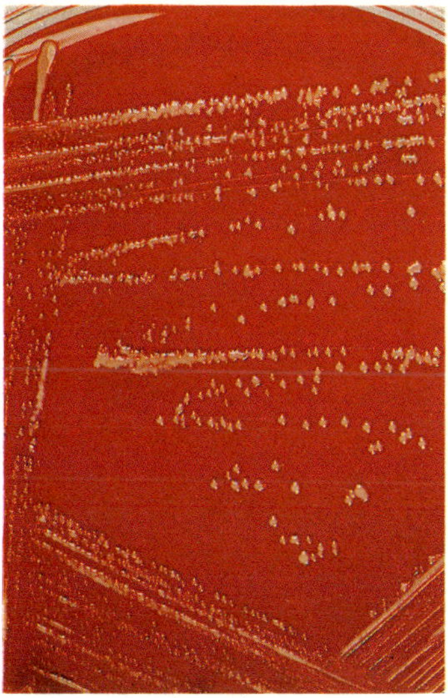 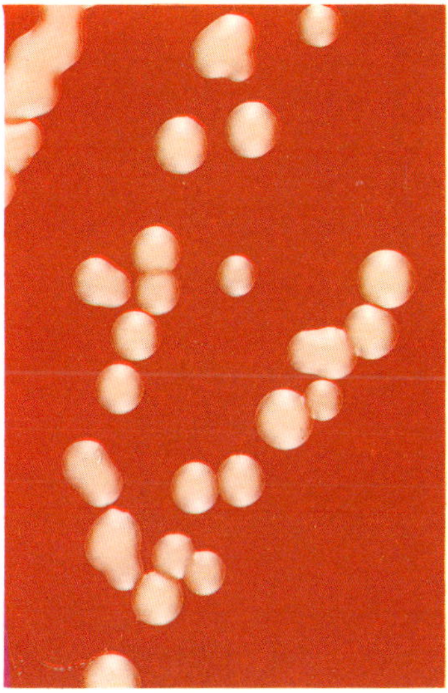 144 B

A, ×1. **B**, ×6.

Figure 144 ***B. insolitus* CCM 2175.** 5 days; 22°C.

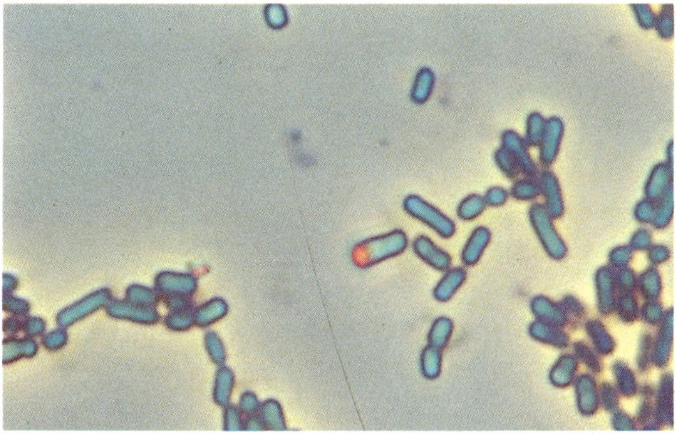 145

Figure 145 Vegetative cells and a sporulating cell of *B. insolitus*. Phase contrast. ×1600.

146

A

B

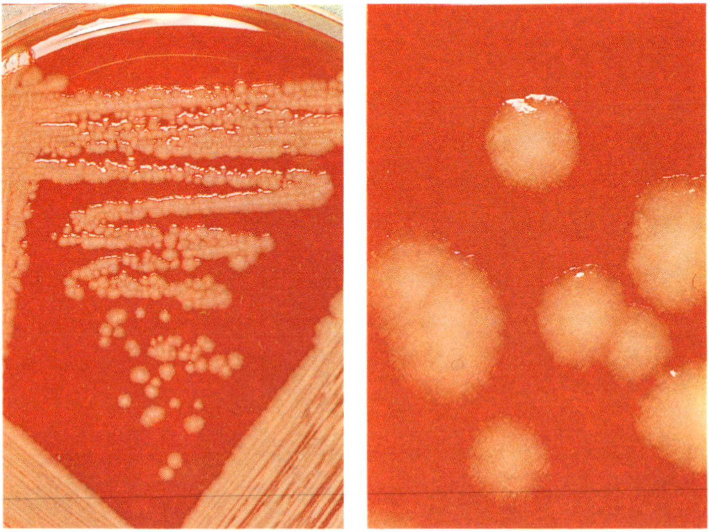

A, ×1. **B**, ×6.

Figure 146 'B. psychrophilus' CCM 2117. 48 hours; 22°C.

'Bacillus psychrophilus'

The nearest relative of 'B. psychrophilus' among the main morphological groups is *B. sphaericus* and it is also closely related to *B. pantothenticus* of the unassigned strains, Subgroup D. Only a handful of strains have been described.

The colonies of strain CCM 2117 (**Figure 146**) measured 1–2 mm in diameter (48 hours; 22°C) and were circular, low convex with an irregular edge. Slight haemolysis was apparent around the fairly moist colonies.

The cells were in the order of $0.85 \times 3.2\,\mu m$ and contained round spores in a subterminal or terminal position slightly swelling the sporangia.

Subgroup E_2

Defining characteristics

Spores oval.
Anaerobic growth positive.
Starch hydrolysis positive.
Grow at 3°C.

Subgroup E_2 species

18. 'B. psychrosaccharolyticus'
19. *B. macquariensis*

‘Bacillus psychrosaccharolyticus’

Unlike the other psychrophiles, ‘B. psychrosaccharolyticus’ is able to grow at temperatures as high as 35°C.

On blood agar, the colonies of strain CCM 2118 (**Figure 147**) showed a variety of forms but were generally convex, mucoidal and creamy-white in colour. As examined at 48 hours (22°C), they had produced slight haemolysis.

The cell size averaged $0.9 \times 2.1\,\mu m$ and the spores were large (relative to the vegetative cell), central and swelled the sporangia (**Figure 148**).

Gibson and Gordon (1974) reported that the cells are distinctly pleomorphic, varying from coccal to elongate and that non-sporulating cells may swell to pear-shaped bodies up to $2\,\mu m$ in diameter. The variety of cell shapes and sizes is visible in **Figure 148** and the body arrowed is probably one of the swollen vegetative cells.

147

A 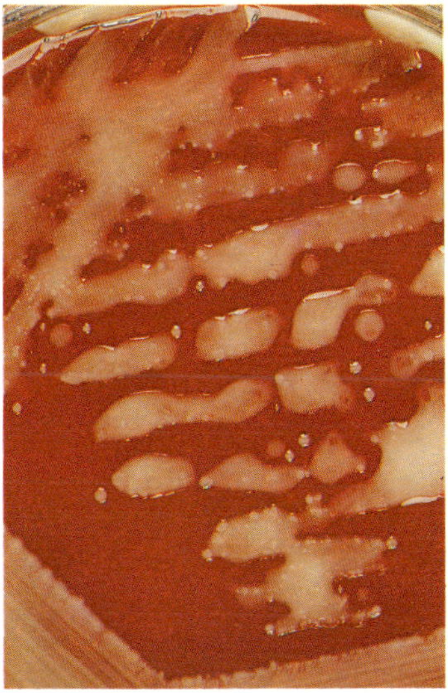 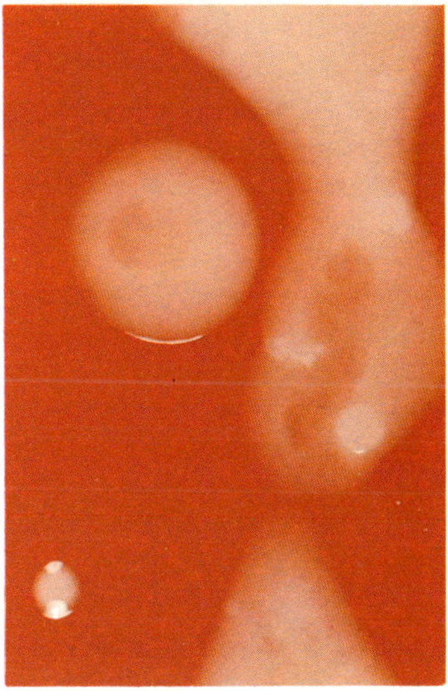 B

A, ×1. **B**, ×6.

Figure 147 'B. psychrosaccharolyticus' CCM 2118. 48 hours; 22°C.

148

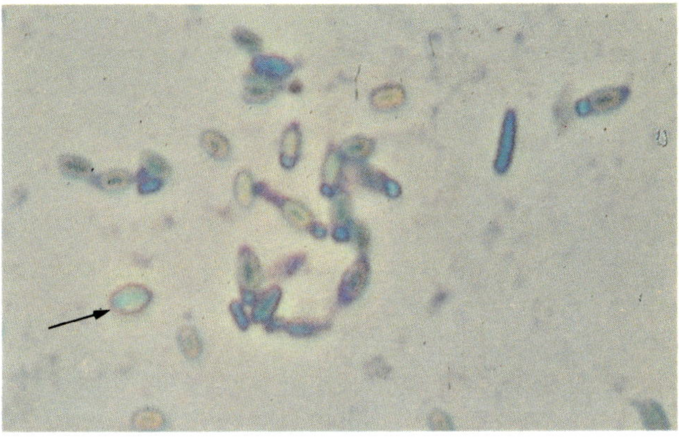

Figure 148 Vegetative cells and spores of 'B. psychrosaccharolyticus'. Phase contrast. ×1300.

Bacillus macquariensis

The three strains in the NCTC and ATCC collections were isolated from the subantarctic soil of Macquarie Island – hence the name given to the species.

The two strains examined (**Figures 149** and **150**) produced similar colonial morphologies on blood agar. The colonies were circular, convex with a slightly irregular edge and measured 0.5–1mm in diameter (48 hours; 22°C); they were creamy-white, moist and butyrous. Slight haemolysis was apparent when the cultures were held a further 2–3 days.

The vegetative cell size was of the order of $0.6 \times 3.25\,\mu m$. The spores were oval and occupied a subterminal or terminal position, swelling the sporangia.

The closest relative of *B. macquariensis* among the main morphological groups appears to be *B. circulans*.

149

A B

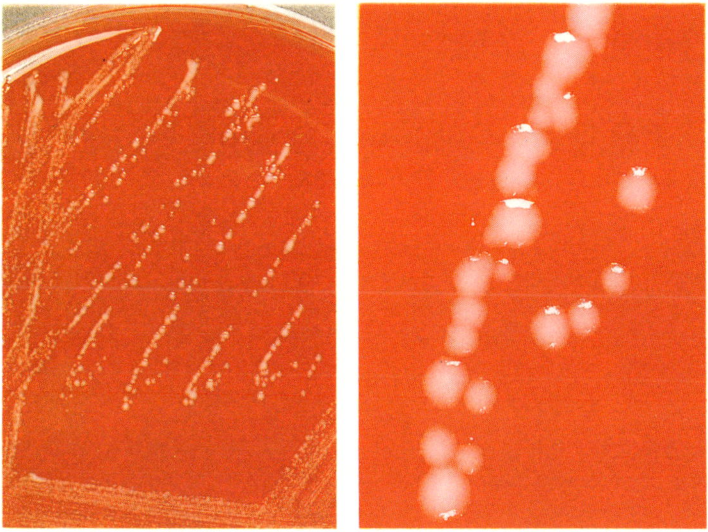

A, ×1. **B**, ×6.

Figure 149 ***B. macquariensis*** **NCTC 10420/ATCC 23465.** 48 hours; 22°C.

150

A B

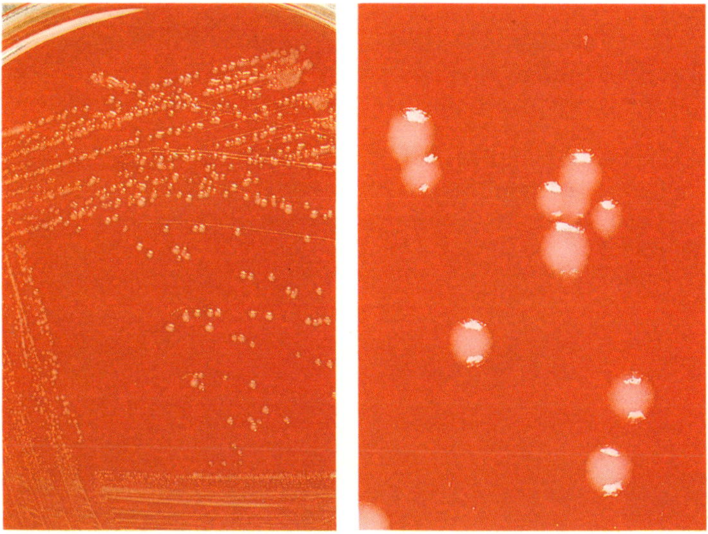

A, ×1. **B**, ×6.

Figure 150 ***B. macquariensis*** **NCTC 10421/ATCC 23466.** 48 hours; 22°C.

Appendix 1

Bacillus species as pathogens

To many, the only accepted pathogen among the *Bacillus* species is *B. anthracis*, the agent of anthrax. Indeed, many medical microbiologists might be hard pressed to name half-a-dozen other *Bacillus* species let alone recognize any of them as pathogens. Even anthrax tends to be regarded now in many parts of the world as of academic, historical or at least of minor significance since it was found that it could be effectively controlled in animals by vaccination, effectively treated in man with penicillin and also since interest in it as a possible germ warfare agent waned. The paper of Turner (1980), however, is a reminder that even today circumstances can occur which inhibit the essentially simple control and treatment and that, under such circumstances, anthrax can still be the scourge it always was. During the insurgence period referred to in that paper, attempts at veterinary vaccination were hampered by ambushes; the result was a large outbreak of bovine anthrax which led to the massive human outbreak.

Other *Bacillus* species have been implicated to one extent or another in infections. *B. cereus* stands out particularly in this respect and has been associated with a wide range of infections, often severe in nature, and two distinct types of food poisoning. *B. subtilis* and *B. licheniformis* have also been implicated in food poisoning and a number of other infections such as bacteraemia and septicaemia, wound infection, respiratory infections, ophthalmitis, peritonitis and, in the veterinary field, bovine mastitis. References to these infections are reviewed by Gilbert *et al* (1981) and Norris *et al* (1981).

Other *Bacillus* species reported on one or more occasions in the literature to have caused or been suspected of causing infections are *B. brevis*, *B. circulans*, *B. coagulans*, *B. macerans*, *B. megaterium*, *B. pumilus*, *B. sphaericus* and *B. thuringiensis*. References to these reports have been given in the respective sections on these species. Gordon *et al* (1973) mention that *B. alvei* has also been isolated from patients in a small number of cases but the details of these were not published.

While, as said in the introduction, it is not our intent to overplay the potential pathogenic roles of *Bacillus* species, there has been a definite recent increase of clinical interest in the genus and the realization in an increasing number of laboratories that these organisms should not be dismissed so readily as unimportant contaminants. This seems to us a trend in the correct direction; while by far the greater majority of isolates undoubtedly are contaminants, we believe that, in laboratories throughout the world, many *Bacillus* species are being discarded daily which were, in fact, of unappreciated relevance to the infections from which they were isolated.

Anthrax

The disease can take a number of forms: intestinal, pulmonary, cutaneous and occasional other conditions such as meningitis.

The intestinal form (not infrequently fatal) arises from ingestion of the organism, often on contaminated meat from animals that have died of the disease. While any of the forms may occur in animals, the majority of cases of infection in animals results from grazing contaminated pasture or ingestion of contaminated fodder.

The pulmonary form results from inhalation of the spores – for example, in dust from contaminated hides of animals that had anthrax. In the pulmonary form (very often fatal), there is characteristically no pneumonia. Mediastinal phagocytes carry the organisms to the regional lymph nodes and from there the disease becomes disseminated systemically.

The cutaneous form, as seen comparatively often on the hands, arms or neck, can frequently be associated with handling or carrying hides or carcasses of animals that had been infected in life. Biting flies are suspected of being not infrequent vectors of cutaneous anthrax (Turner, 1980).

The marked characteristic of cutaneous anthrax (which is seldom fatal) is an accompanying gross oedema often spreading some distance from the lesion itself.

151

A

B

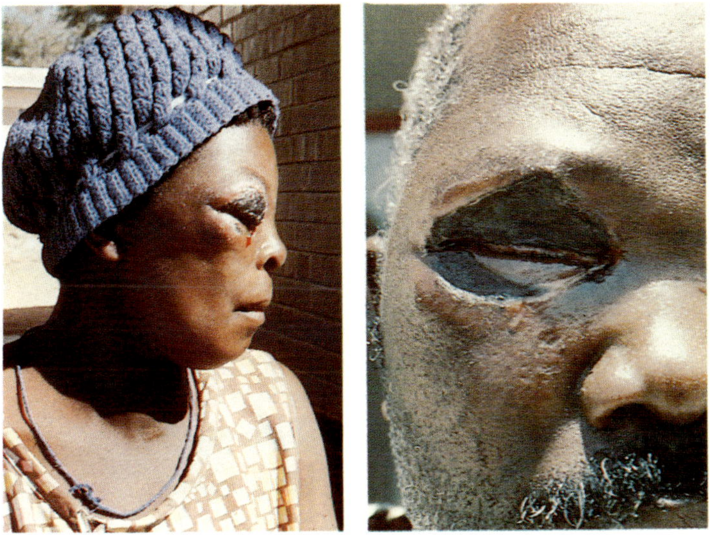

Figure 151 Anthrax infection of the eye.
A, Acute stage. (Kindly supplied by Dr A.O. Pugh, Provincial Medical Officer of Health, Bulawayo, Zimbabwe.)
B, Convalescent stage. (Kindly supplied by Squadron Leader J.E. Bland, Medical Officer, RAF Hospital, Ely, Cambridgeshire.)

In eye infections (**Figure 151**), the acute stage is marked by oedema of the eyebrows and surrounding regions of the temples and face. The upper lid is usually the site of the typical black eschars, but the eye itself is characteristically not damaged and recovery of sight is complete in the absence of secondary complications (Dr J.C.A. Davies, personal communication).

152

A

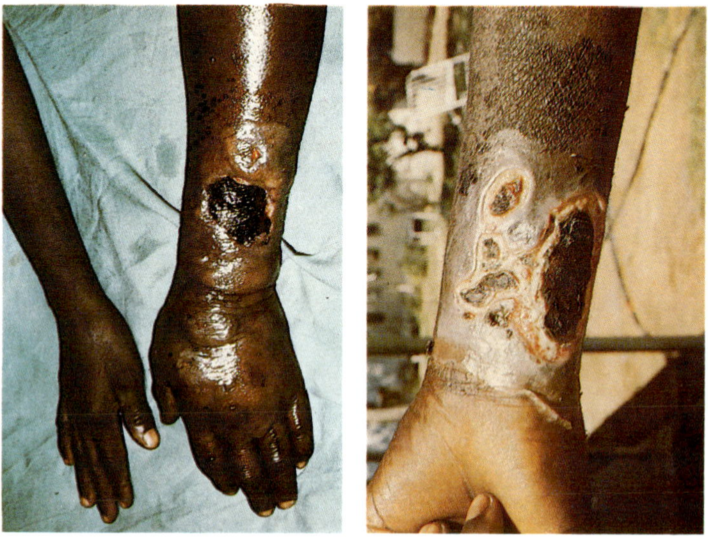

B

Figure 152 Anthrax infection on the wrist.
A, Acute stage. (Kindly supplied by Dr J.C.A. Davies, formerly Medical Officer of Health, Salisbury, Zimbabwe.)
B, Convalescent stage. (Kindly supplied by Squadron Leader J.E. Bland, Medical Officer, RAF Hospital, Ely, Cambridgeshire.)

In **Figure 152**, the black eschars typical of cutaneous anthrax are visible.

The acute stage (**Figure 152A**) is marked by massive oedema which often spreads up the entire arm and down part of the trunk. The organisms are numerous in the lesion and on the eschar and are readily isolated with a swab.

In **Figure 152B**, an older lesion, the oedema has receded leaving the skin scaly and sloughing above the healing lesions. Multiple lesions such as seen here are not common.

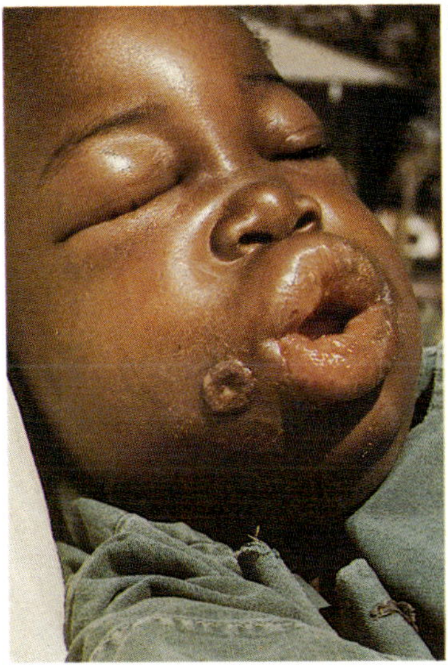

153

Figure 153 Anthrax infection of the face. (Kindly supplied by Dr J.C.A. Davies, formerly Medical Officer of Health, Salisbury, Zimbabwe.)

The lesion on this child's face (**Figure 153**) probably results from the infected bite of a biting fly, the fly having carried the organism on its proboscis. The eye infections in **Figure 151** may also be examples of fly-borne infections. The higher incidence of cutaneous anthrax in endemic regions during hotter months may be attributable, at least in part, to higher fly populations (Dr J.C.A. Davies, personal communication).

In **Figure 153**, note the intense oedema of the mouth and lips.

Recovery with cutaneous anthrax is very slow, taking several weeks, as a result of widespread endothelial cell damage in the vicinity of the lesion. There is no permanent damage; after recovery, normal functions return to the area. Penicillin makes no difference to the course of the lesion but prevents further complications.

***Anthrax vaccines* (Figure 154)**

In regions of the world where anthrax is endemic, the first approach to control is annual vaccination of livestock.

There are two kinds of vaccines available for animals: the Sterne-type using spores of a non-encapsulated strain of *B. anthracis* and the Pasteur-type which is made up of spores from encapsulated strains. There are several Pasteur-type vaccines designated in accordance with their degree of virulence.

Figure 154 shows the Sterne-type, developed in South Africa and used with excellent results there and in Britain and other countries. It is a live spore suspension of an avirulent mutant strain obtained by growing a virulent strain on serum-rich medium in the presence of carbon dioxide and selecting single rough colonies. These are then cultured on a sporulation medium and the resulting spore suspension is suspended in a mixture of glycerol and saponin. Quality control involves testing the efficacy of the vaccine's protective activity in guinea pigs. The method and the strain, $34F_2$, for producing this vaccine are those of Sterne (1939).

One ml of the vaccine is administered subcutaneously regardless of species, breed or age of the animal (the vaccine is listed as being for the immunization of cattle, horses, mules, donkeys, sheep, goats, pigs and camels). The glycerol and saponin cause a local tissue reaction at the site of injection which permits the spores to germinate and the resulting vegetative cells to multiply and spread throughout the body leading to development of effective immunity.

154

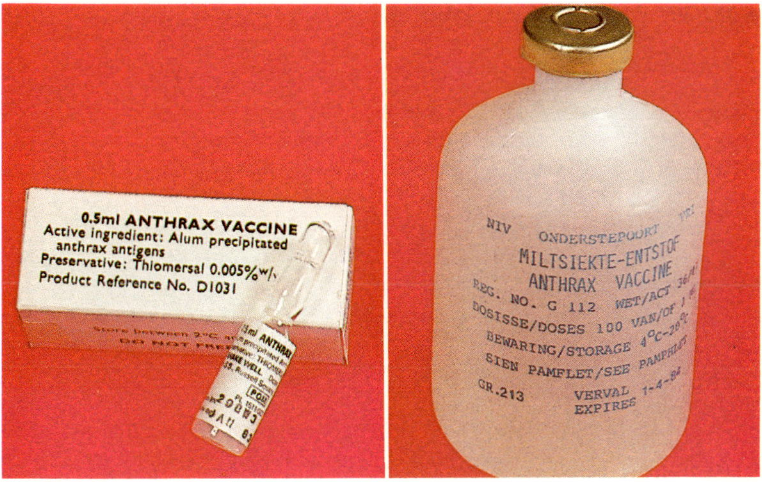

Figure 154 Anthrax vaccines.

Left: vaccine for human use. (Kindly supplied by Dr J. Melling, Vaccine Research & Production Laboratory, PHLS–CAMR, Porton Down, Wiltshire.)

Right: vaccine for animal use. (Kindly supplied by Dr C.M. Cameron, Veterinary Research Institute, Onderstepoort, South Africa.)

A vaccine for human use is also available for veterinarians, hide-sorters, certain laboratory personnel and others at occupational risk. This consists of the sterile filtrate of a culture of the Sterne strain $34F_2$ (NCTC 8234) grown essentially by the method of Belton and Strange (1954). The filtrate contains the 'protective antigen' of *B. anthracis*; this is concentrated by precipitation with 0.1% aluminium potassium sulphate (alum).

The course of immunization consists of three doses of 0.5 ml administered intramuscularly at intervals of 3 weeks with a fourth dose at 6 months and reinforcing doses annually.

Bacillus cereus infections

The growing number of reports of infections caused by *B. cereus* has been reviewed elsewhere (Norris *et al*, 1981; Turnbull, 1981). The bacterium has been associated from as early as 1898 with abscess formation, bacteraemia and septicaemia, cellulitis, ear and eye infections, endocarditis, gastroenteritis, meningitis, kidney and urinary tract infections and wound infections, sometimes gangrenous. Frequently these infections have been severe, occasionally fatal.

B. cereus is also known in the veterinary field, particularly in relation to bovine mastitis.

These infections are represented by a few examples in **Figures 155–158**.

With respect to gastroenteritis, *B. cereus* has been associated with two distinct types of food poisoning: (1) a 'diarrhoeal type' characterized by symptoms predominantly of diarrhoea and abdominal cramps 8–16 hours after eating the implicated food and (2) an 'emetic type' in which the predominant symptom is vomiting 1–5 hours after consumption of the incriminated food. The two syndromes have been attributed to two unrelated toxins produced under different conditions of mishandling of foods (Gilbert, 1979; Gilbert *et al*, 1981; Turnbull, 1981).

B. cereus endocarditis (Figure 155)

This case was reported by Block *et al* (1978). The patient, a 51-year-old woman with chronic rheumatic heart disease, had undergone prosthetic mitral valve replacement. She became febrile on the 17th postoperative day. Ten blood cultures yielded *B. cereus*. Despite treatment with tobramycin and chloramphenicol, she died 72 hours later.

At postmortem, the infective endocarditis shown in **Figure 155** was seen. The large vegetations (arrows) consisted histologically of neutrophil infiltrate consistent with endocarditis. No bacteria were actually observed and unfortunately, due to a delay in the autopsy, no specimens for culture were taken.

155

A

B

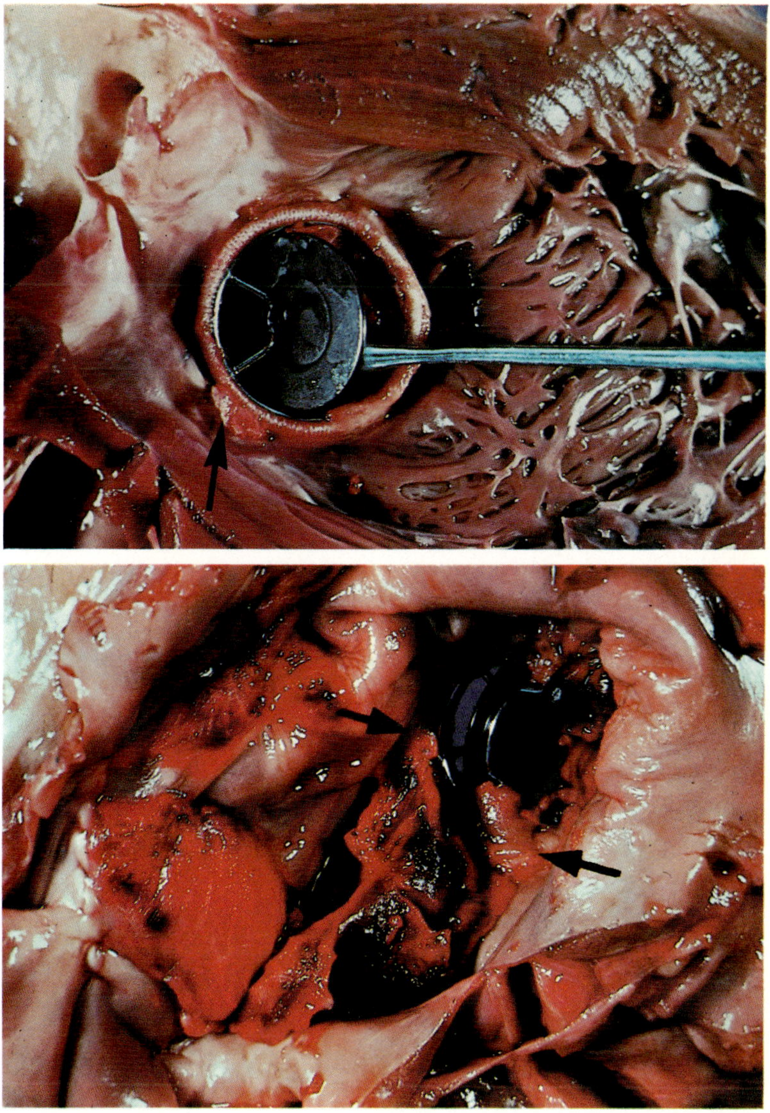

Figure 155 A case of *B. cereus* endocarditis. (Kindly supplied by Dr C.S. Block and Professor H.J. Koornhof, Department of Microbiology, South African Institute for Medical Research, Johannesburg.)

***B. cereus* panophthalmitis (Figure 156)**

B. cereus is a highly significant agent of ocular infections following foreign body eye injuries (O'Day *et al*, 1981). The infection is usually in the form of a keratitis, endophthalmitis or panophalmitis. The course can be swift and devastating and the sight following fulminating panophthalmitis is unlikely to be saved.

The case shown in **Figure 156A** was reported by Young *et al* (1980). A 23-year-old man presented with pain in the left eye attributed to tobacco ash that had flown into his eye the previous day. The next day he was mildly febrile and the eye was swollen and chemotic with marked periorbital oedema (visible in **Figure 156A**). His WBC count was elevated. An aspirate of the vitreous revealed Gram-positive bacilli which, on culture, were identified as *B. cereus*. Despite clindamycin therapy, there was rapid progression of infection and two days later, corneal perforation occurred requiring enucleation.

The case shown in **Figure 156B** was reported by O'Day *et al* (1981). A 7-year-old boy was striking a manure spreader with a hammer when a metal fragment penetrated his left eye. The following day, the eye became proptosed and there was marked chemosis of the conjunctiva; after a further 24 hours, he was febrile with an elevated WBC count. The eye now had the well-demarcated corneal ring abscess around the periphery of a uniformly oedematous cornea (**Figure 156B**). Numerous Gram-positive bacilli were visible in the vitreous aspirate which, on culture, proved to be *B. cereus*. The patient recovered slowly with systemic and periocular therapy, but vision was permanently lost.

There have been scattered reports of eye infections due to other *Bacillus* species – *B. brevis*, *B. coagulans*, *B. laterosporus* and *B. megaterium*. The references to these reports are given in the sections on the respective species. A *B. licheniformis* infection resulting in corneal ulceration has also been reported (Tabbara and Tarabay, 1979).

156

A

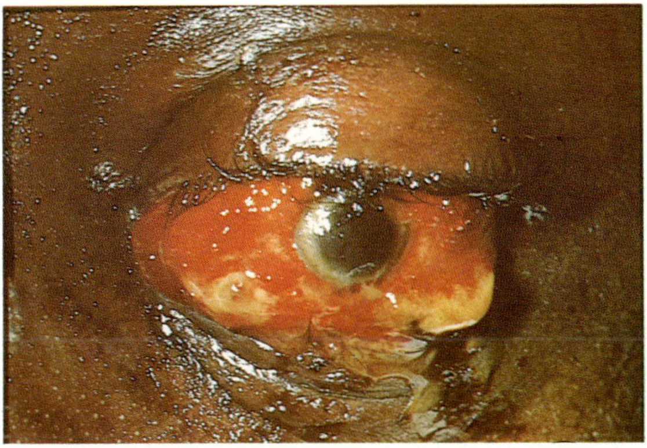

B

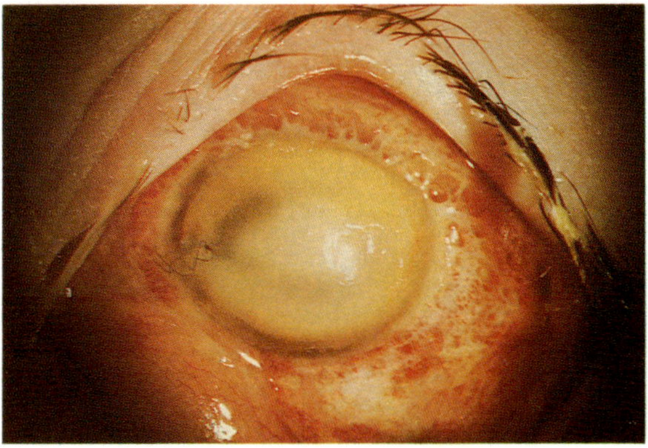

Figure 156 ***B. cereus* panophthalmitis.**

A, Case 1: second day of infection showing marked haemorrhage, chemosis and periorbital oedema. (Kindly supplied by Dr E.J. Young. Reproduced, with permission from the *Archives of Internal Medicine*, April 1980, Volume 140, pages 559–560. Copyright 1980, American Medical Association.)

B, Case 2: corneal ring abscess in a patient with *B. cereus* panophthalmitis following a penetrating intraocular injury. (Kindly supplied by Dr D.M. O'Day, Department of Ophthalmology, School of Medicine, Vanderbilt University, Nashville, Tennessee 37232.)

Mastitis

B. cereus is among a number of bacteria that can cause mastitis in bovines and other animals. This may range from mild to gangrenous and rapidly fatal (Jones and Turnbull, 1981). There are numerous other reports of *B. cereus* bovine mastitis in the literature.

Two cases of bovine mastitis are seen in **Figure 157**. In **Figure 157A**, some darkening of the skin and swelling of the front right quarter is visible. The other quarters are slack and flaccid indicating the cow has ceased to produce milk.

In **Figure 157B**, purulent discharge through the udder wall has caused matting of the hair.

B. licheniformis and *B. subtilis* have also been associated with bovine mastitis (Mr C.L. Wright, personal communication).

A

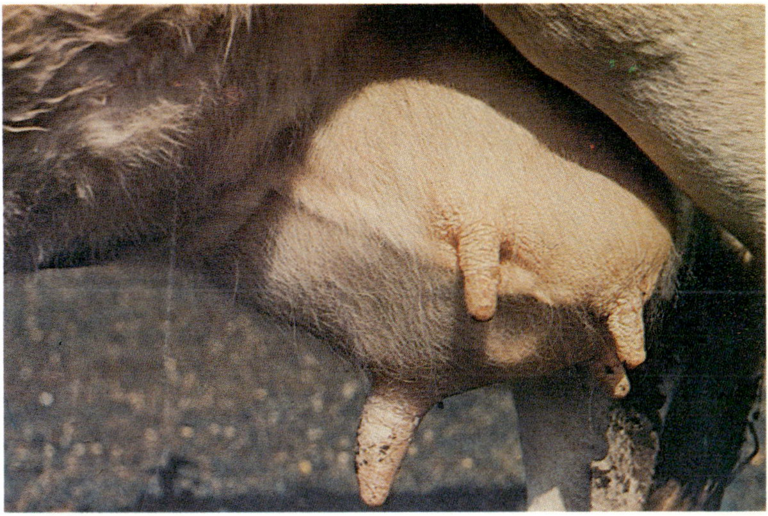

B

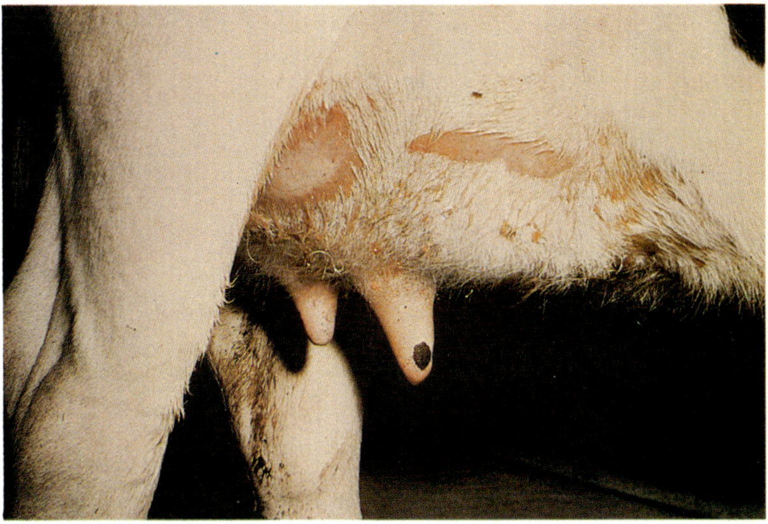

Figure 157 Bovine mastitis. (**A**, kindly supplied by Mr C.L. Wright, MRCVS, Auchincruive V.I. Centre, Scotland. **B**, kindly supplied by Mr T.O. Jones, MRCVS, V.I. Centre, Loughborough, Leics.)

Mastitic milk

In a low grade or early stage of infection, milking cows with *B. cereus* mastitis results in milk tinged pink with blood. (**Figure 158A–RF**.)

In later stages of fulminant infection, the milk is replaced by a sero-sanguinous discharge, port-wine to rusty brown in colour (**Figure 158A–RH** and **Figure 158B**). 'Clinicians . . . suspect *B. cereus* on sight when confronted with these colours. Occasionally the same colours are seen with acute *Escherichia coli*, *Staphylococcus aureus* and the rare *Clostridium perfringens* infections' (K.G. Johnston, personal communication; see also Johnston, 1981).

The organism is readily isolated in profuse, probably pure culture from such 'milk'.

A

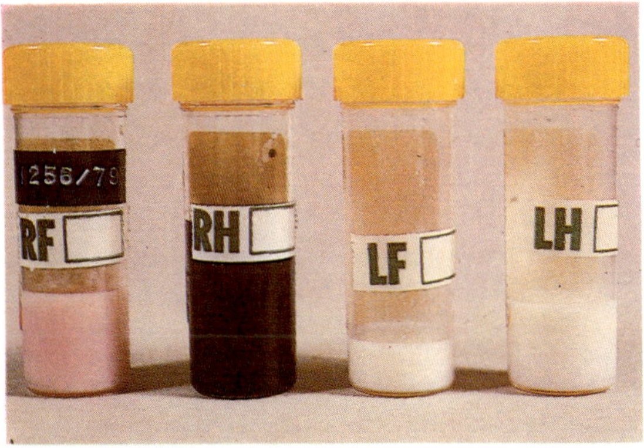

A Mastitic milk.

B

Figure 158 B Serosanguinous exudate from mastitic teats. (Kindly supplied by Dr K.G. Johnston, MACVSc, Senior Clinical Pathologist, University of Sydney, Rural Veterinary Centre, Werombi Road, Camden, NSW 2570, Australia.)

Pathogenesis of *B. cereus* infections

It has been known for many years by some, if not appreciated by all microbiologists, that *B. cereus* is capable of producing serious infections (Turnbull, 1981). There has naturally followed some curiosity as to the mode of pathogenic action by which it does so.

The organism has been known for some time to produce a host of 'aggressins' – cytolysins, tissue-destroying enzymes, nucleic acid-destroying enzymes, protein coagulating or destroying enzymes and so on. These can be readily detected *in vitro* but their *in vivo* significance can only be guessed at. For example, it is known that haemolysin activities are rapidly neutralized by serum and so, presumably, when produced during an infection, they are rapidly neutralized in the blood stream. The same may occur with other aggressins also.

There appears to be a correlation between severity of infection and the strength of the reaction obtained in the skin test shown in **Figure 159**. In this, 0.05 ml of cell-free culture filtrate (0.45 μm membrane filter) from a 6½-hour brain–heart infusion broth culture of the strain is injected intradermally at 0 hours into the shaved skin of an adult rabbit. At 3 hours, 4 ml of 2% Evans blue dye is injected intravenously and, after a further hour, the zones of blueing and, if present, necrosis are measured.

Strains producing the toxin(s) responsible for the skin test reaction give rise to the skin necrosis seen at some of the injection sites in **Figure 159A**. Other strains only produce the toxin(s) relatively weakly.

Work is underway to purify the toxin or toxins concerned so that non-animal methods can be developed to detect and quantitate toxin production by different strains and its/their presence in clinical specimens, foods, etc.

B. licheniformis and *B. subtilis*, which have also been implicated in occasional diarrhoeal-type food poisoning and other infections, do not produce this toxigenic effect. *B. thuringiensis*, which is a very close relative of *B. cereus*, can produce it in the same manner as *B. cereus* (**Figure 159B**).

159

A

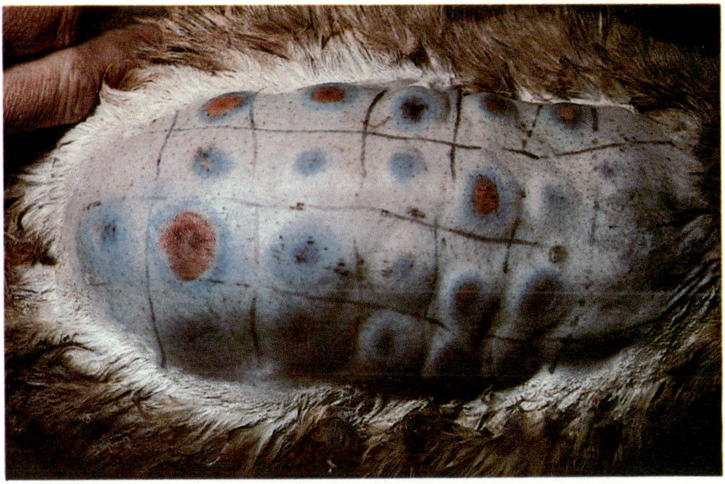

A Skin test toxin of ***B. cereus.***

B

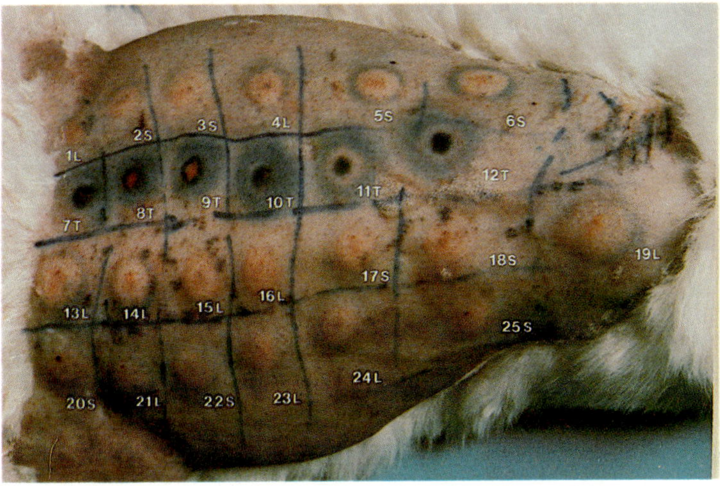

Figure 159 B Skin test using other *Bacillus* species. L = *B. licheniformis*; S = *B. subtilis*; T = *B. thuringiensis*. The numbers refer to different strains being tested.

The zone of blueing in the skin test demonstrated in **Figure 159** is indicative of increased permeability in the blood capillaries of the skin to the blood plasma. This, in the case of *B. cereus* toxin, presumably results from damage to the capillary walls by the toxin. Leakage of plasma out of the blood vessels into the surrounding tissue becomes visible after intravenous injection of the dye into the blood stream.

In histological cross-sections of the skin (**Figure 160**) the effect of the toxin(s) can be seen.

Figure 160A is an unstained cross-section of rabbit skin. Blue dye can be seen to have perfused into the tissue surrounding the site of injection (arrow).

Figure 160B is the same section as **Figure 160A** but stained with haematoxlyin-and-eosin. The intensely staining region around the site of injection marks the zone of necrosis.

A less intense reaction is seen in **Figure 160C** than that seen in **A** and **B**. Necrosis was not apparent but the capillaries were sufficiently damaged for blood to leak out into the tissues (arrows).

Figure 160D is a piece of normal skin for the purposes of comparison with **C**.

It may be that a combination of more than one *B. cereus* metabolite is responsible for these effects of altered vascular permeability and necrosis; at present, it is believed that the same toxin or group of toxins is responsible for producing diarrhoea in the diarrhoeal type of *B. cereus* food poisoning.

160

A

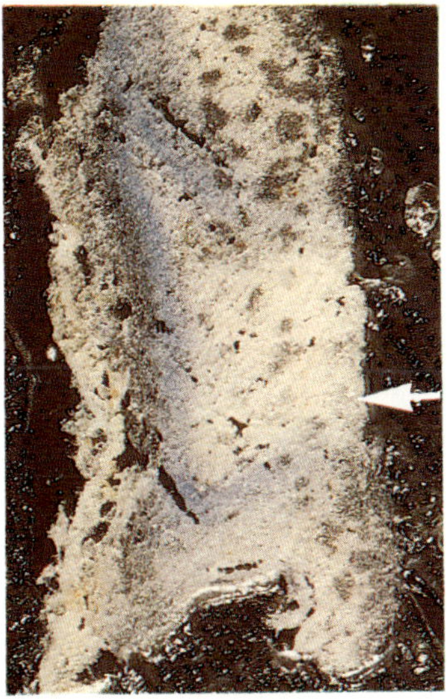

B

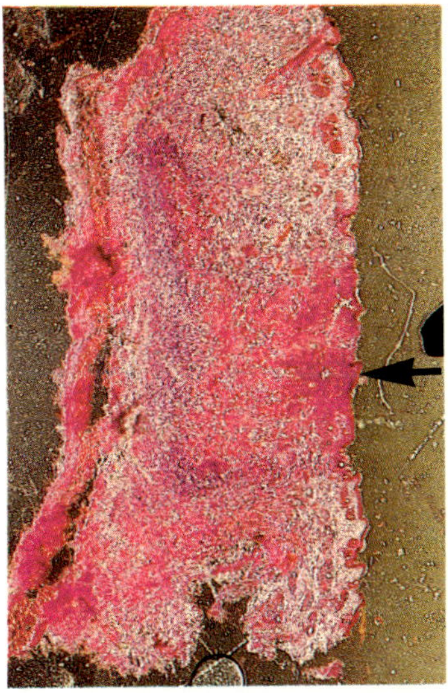

A, × 8. Unstained.

B, × 8. H&E stain.

C

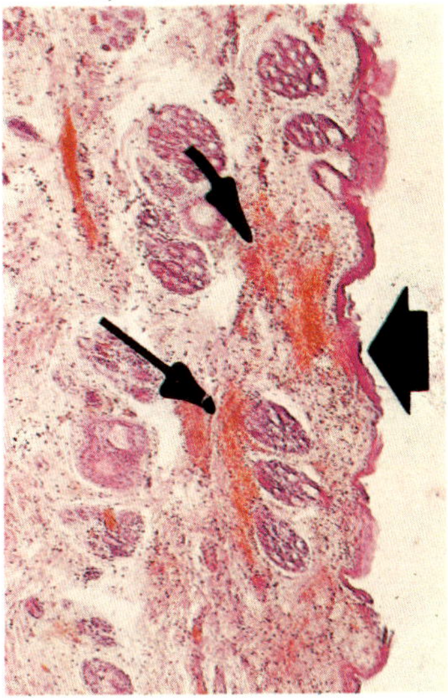

D

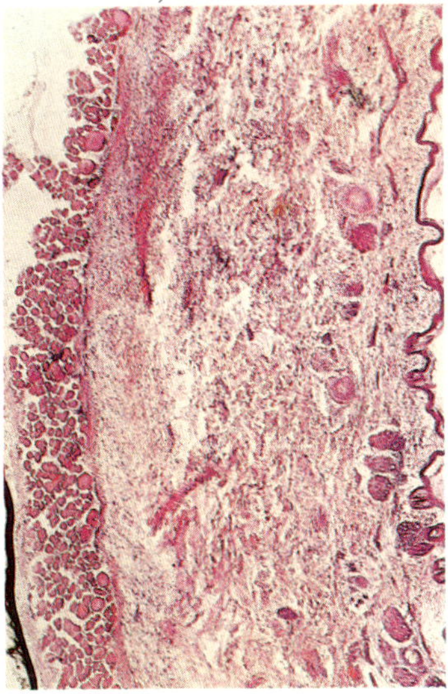

C, × 50. H&E stain.

D, × 24. H&E stain.

Figure 160 Histological cross-sections of rabbit skin after the skin test.

Appendix 2

Identification systems for *Bacillus* species

Only one commercial system, that produced by API Laboratory Products Ltd, is available at the time of writing.

The system is based upon the API 50CHB (*Bacillus*) kit which comprises the API 50CH gallery, containing 49 carbohydrates and one control in 5 strips of 10 cupules, and an ammonium salts inoculating medium (API 50CHEB medium) containing phenol red as indicator.

The API 50CHB is intended to be used in conjunction with the first 12 tests of the API 20E (Enterobacteriaceae) strip, both kits being set up after the organism has been confirmed as *Bacillus* by microscopic observation for endospores and sporangial morphology. Routine studies of vegetative cell morphology and motility are also recommended.

Following inoculation and incubation (48 hours at 30–37°C for most strains), the test results are compared with a table of per cent positive results which is based on the reactions of over 1000 strains. This allows the identification of atypical as well as typical strains.

The API *Bacillus* system should prove of great value to laboratories involved in the identification of large numbers of strains, especially where the species range is wide. The manufacturer claims that the cost of using the kits is comparable to that of the preparation of conventional tube tests for *Bacillus* and the API method has the advantage of convenience, rapidity, and accuracy due to standardization of the test media.

The development of the system is described by Logan and Berkeley (1981).

We are grateful to Dr N.A. Logan, Department of Biological Sciences, Glasgow College of Technology, Cowcaddens Road, Glasgow, for assistance in the preparation of this appendix.

Appendix 3

Uses of *Bacillus* species

It is clear from the information given in the American Type Culture Collection Catalogue of Strains I (1982) that *Bacillus* species have been or are being put to a very wide range of important medical, agricultural, pharmaceutical or other industrial uses throughout the world.

Tables 7–11, compiled from the information in the ATCC catalogue, illustrate this wide variety of uses. 'Production', as used in the titles of **Tables 8**, **9** and **10**, is taken from the information given in the catalogue as indicating 'commercial' or 'industrial' production.

Table 7 Assay of antibiotics: *Bacillus* strains recommended.

Bacillus species	ATCC No.	Antibiotic assay
B. cereus	10702	Aureomycin
	11778	Vancomycin, tetracyclines
	13061	Penicillin index
	19637	Flavomycin
	27348	Mixtures of ß-lactam antibiotics
B. circulans	9966	Streptomycin
B. megaterium	9885	Streptomycin
	25833	Cycloserine
B. pumilus	14884	Neomycin
B. stearothermophilus	10149	Presence of antibiotics in milk, body fluids, etc.
B. subtilis	6633	28 antibiotics listed
	9524	Streptomycin, penicillin
	11838	Viomycin
	12432	Novobiocin

Table 8 Production of antibiotics.

Bacillus species	ATCC No.	Antibiotic
B. brevis	8185, 10068	Gramicidin, tyrocidin
	9999	Gramicidin S
	21991	342–14–I
B. cereus	21929	60–6
B. circulans	21557/8	Butirosin
	21656	EM-49 complex
	21828	Bu-1880
	21820/1/2	4-deoxyambutyrosin
B. licheniformis	10716	Bacitracin
	11945/6	Bacitracin
	21552	Proticin
B. polymyxa	10401	Polymyxin
	21830	Gatavalin, jolipeptin, colistin
	31037	BN-109
Bacillus spp. (unnamed)	21932/3	Xylostasin
B. subtilis	10774	Bacillomycin B
	14593	Bacitracin
	14807	Bacillin

Table 9 Production of enzymes.

Bacillus species	ATCC No.	Enzyme(s)
B. alcalophilus	21522	Alkaline protease
**B. amyloliquefaciens*	23842–5	Amylase, protease, RNase
B. cereus	13061	Penicillinase (penicillin index)
	14893	Bacteriolytic enzyme
	21182	Urate oxidase
	21634	Polynucleotide phosphorylase

*Not covered in the atlas; closely related to *B. subtilis*.

Table 9 Continued

Bacillus species	ATCC No.	Enzyme(s)
B. cereus	21768–72	∝-amylase
	27348	Broad-spectrum ß-lactamase (for antibiotic assay)
B. circulans	15329	Neuraminidase
	21367	Enzyme lytic on yeast cell walls
	21590	Peptidoglutaminase
B. coagulans	21366	Enzyme lytic on yeast cell walls
B. licheniformis	12713	Pentosanases (removal of pentosan gum from wheat flour)
	21424	Alkaline protease
	25972	Penicillinase (prodigious production)
	27811	Thermostable ∝-amylase
B. polymyxa	12712	Pentosanases
	21551	Pectic acid trans-eliminase (degrading plant material)
	21993	Neutral protease
	25901	Aspartokinase
Bacillus spp. (unnamed)	31084	DNase
	21536/7	Alkaline protease
	21592–6	Alkaline amylase
	21832/3	Alkaline cellulase
B. stearothermophilus	21365	Enzyme lytic on yeast cell walls
B. subtilis	6051a, 21556	∝-amylase, protease
	12711	Pentosanases
	21183	Urate oxidase
	21228	Alkaline protease
	21415–8	Alkaline detergent-resistant protease

Table 10 Production of organic compounds

Bacillus species	ATCC No.	Production of
B. cereus	31101–3	Maltose
B. circulans	13403	L-glutamic acid
B. licheniformis	9945	D-glutamic acid
	9945a	Glutamyl polypeptide
	21037–9	5′-inosinic acid and inosine
	21552	'Proticin'
	21667	Citric acid
	27811	Thermostable ∝-amylase
B. megaterium	13402, 15177, 15781	L-glutamic acid
	15406/7	Xanthosine
	15117/8	5-amino-imidazolecarboxamide riboside
	15127/8	Purine derivatives
	15450/1	Fructose
	19135–7	5′-inosinic acid
	19160/1	5′-inosinic acid and inosine
	19218	Guanosine and 5′-guanylic acid
	21181	Urate oxidase
	21209	L-lysine
	21603	L-histidine
	21737/8	A sucrose-starch sweetener
	21916	2-keto-L-gulonic acid
'B. pulvifaciens'	21112	D-pantoic acid
B. pumilus	15477	5′-amino-4-imidazole-N-succino-carboxamide riboside
	19164	5′-inosinic acid and inosine
	19182	Xanthylic acid
	19217/9/20	Guanosine and 5′-guanylic acid
	21005–8	Inosine
	21143	3-amino-3-deoxy-D-glucose
	Several	D-ribose

other than antibiotics and enzymes.

Bacillus species	ATCC No.	Production of
Bacillus spp. (unnamed)	13062	L-glutamic acid
	15192	Heteropolysaccharide 'thickener'
	19385	Protein and vitamin food and feed supplement
	Several	Cylodextrins
	21616	Adenosine
	21617	Xanthosine
	21618	Guanosine
	27860	L-sorbosone
B. sphaericus	13805	1,4-diene-3-ketosteroids
B. subtilis	13933	Aterrimin and feed supplements
	13952–6	Inosine
	14617/8	Inosinic acid and adenylic acid
	15039–44	Xanthosine
	15115/6	5-amino-imidazolecarboxamide riboside
	15129	Purine derivatives
	15181–4	Citrulline
	15811	5′-nucleotides
	19062	5-fluorouracil and 6-mercaptopurine ribosides
	19162/3	Inosinic acid and inosine
	19221	Guanosine and inosine
	19549/50	Diaminopimelic acid
	21331	'Surfactin'
	21336, 21777/8	L-tryptophan
	Several	D-ribose
	21556	Brewer's wort (∝-amylase and protease)
	21610	Citric acid
	21733	5-OH-L-tryptophan
	31002–4	L-arginine

Table 11 Other uses of *Bacillus* species.

Bacillus species	ATCC No.	Use
B. cereus	21281/2	Insecticide for mosquito larvae
B. coagulans	12245	Assay of folic acid
B. megaterium	10778	Production of vitamin B_{12} and riboflavin
	13368	Introduction of 15-OH group into steroids
	13639	Biosynthesis of cobalamin (nucleus of vitamin B_{12})
	25848	Assay of aflatoxin
B. pumilus	27142	Testing radiation resistance of spores
B. stearothermophilus	7953	Steam sterilization control
B. subtilis	6051	Blood screening for phenylketonuria
	6533	Assay of hexachlorophene
	19659	Sporicidal tests
	21663	Increasing fermentation yields
	21697	Degradation of nitriles and cyanides in waste water effluent
B. popilliae, *B. sphaericus*, *B. thuringiensis*	Not applicable	Insect pest control

Microbial pesticides

Reference to the pathogenic nature of *B. popilliae*, *B. sphaericus* and *B. thuringiensis* for various insect species was made in the respective sections of the main text and it was mentioned there that advantage was being taken of this in the biological control of a variety of insect pests.

B. popilliae was the first to be registered for this type of use in 1950 but its limitations lie in the fact that it can only be grown in living insects. However, it spreads effectively and gives permanent control of Japanese beetle in regions with a sufficiently warm climate (Klein, 1981).

B. thuringiensis, first registered in 1960, is the most extensively used for insecticides; by 1979, more than 7000 tons of commercial insecticides, comprised of viable spores of *B. thuringiensis* mixed with its parasporal crystals, had been used worldwide – a significant proportion on food crops (Burges, 1981).

Table 12 shows some registered uses of *B. thuringiensis* insecticides.

Interest in *B. sphaericus* for mosquito control has increased recently. Results of trials have suggested that *B. sphaericus* larvicides may compete favourably with chemical larvicides, particularly at sewage treatment plants and in highly polluted water in which the bacteria may reproduce saprophytically to give extended mosquito control (Singer, 1981; Burges, 1982).

These products are subjected to exhaustive tests for their safety in relation to human and animal exposure.

Table 12 Some registered uses of *B. thuringiensis* insecticides in the USA (from Falcon, 1971).

Insect	Crop
Alfalfa caterpillar	Alfalfa
Artichoke plume moth	Artichokes
Boll worm	Cotton
Cabbage looper	Cabbage, cauliflower, celery, beans and broccoli
Diamond back moth	Cabbage
European corn borer	Maize
Fruit tree leafroller	Oranges
Grape leaf folder	Grapes
Imported cabbage worm	Cabbage, cauliflower and broccoli
Orange dog	Oranges
Tobacco budworm (hornworm)	Tobacco
Tomato hornworm	Tomatoes
California oakworm Fall webworm Fall cankerworm Great Basin tent caterpillar Gypsy moth Linden looper Salt-marsh caterpillar Spring cankerworm Wintermoth	Forests, shade trees and ornamentals

B. stearothermophilus in a commercial antibiotic assay

As can be seen in **Table 7** a number of strains among the *Bacillus* species are recommended for use in assays of various antibiotics. Mostly, these relate to specific tests concerned with particular antibiotics.

B. stearothermophilus finds use in an interesting and rather more general assay for antibacterial drugs and antibiotic residues in milk and dairy products and body fluids.

As was mentioned in the section on *B. stearothermophilus* in the main text, the advantages of using this species in these types of tests are (1) its sensitivity to a wide range of antibiotics, (2) the rapidity with which the organism grows at 65°C – and therefore the rapidity with which the test can be read and (3) in the case of milk and dairy products, the lack of interference in the test by other organisms that may be present but which cannot grow at 65°C.

The commercial product Thermocult(R) takes advantage of these factors. It consists of a ready-poured nutrient agar plate containing a pH indicator and spores of *B. stearothermophilus* var. *calidolactis*.

The paper discs are immersed in the fluids being tested for the presence of antibiotics and placed together with a reference control disc (containing 0.01 u/ml of penicillin G) on the plate. On incubation at 65°C, the *B. stearothermophilus* spores vegetate and growth is such that the plate can be read at 3 hours.

B. cereus in an anti-diarrhoeal formulation

Bactisubtil$^{(R)}$ is a product prescribed in France and certain other countries of Europe for the treatment of diarrhoea, enteritis and colitis, particularly, it appears, as induced by antibiotic therapy.

It is produced in liquid form in vials and dry in capsules. In both cases it consists of viable spores of Institut Pasteur culture IP 5832 – 500 million spores per ampoule in liquid form and 1000 million spores per capsule in dry form. The dose is 3–6 ampoules or capsules per day for adults and half this for infants and children.

Bacillus IP 5832 was classified in 1949 (before order had been brought to the taxonomy of *Bacillus* species) as *B. subtilis* – hence the product name. Subsequently it was identified as *B. cereus*. The strain as used in Bactisubtil was rendered resistant to several antibiotics; interestingly, however, it is susceptible to penicillin – which is unusual for *B. cereus*.

The 'replacement flora' principle embodied in this product is not especially unusual; there are several *Lactobacillus* preparations on markets throughout the world sold for the same types of ailments and based on some of the same principles. In the manufacturer's leaflet accompanying Bactisubtil$^{(R)}$, it is suggested that production of enzymes is involved in the effectiveness of the product.

B. subtilis for removal of effluent slurry

An X-ray mutated strain of *B. subtilis* is one of the active ingredients of the product Hyzyme$^{(R)}$ developed for biological decomposition of sludge in effluent pipes, septic tanks or cesspools.

Enzymes produced by the bacteria, together with added protease, amylase and lipase, break down insoluble proteins, starches, cellulose and fats which make up the unwanted sludge.

Examples of advantages over chemical treatments are persistence of effect through continued multiplication of the *B. subtilis*, lack of corrosiveness to metal parts and lack of residual chemicals in the treated effluent.

The product has, of course, been extensively tested for its safety as related to the exposure of man or animals (including fish) in the event of the effluent reaching streams or rivers or being used as fertilizer.

Key to culture collections

ATCC = American Type Culture Collection, 12301 Parklawn Drive, Rockville, Maryland, 20852, USA.
CCEB = Culture Collection of Entomogenous Bacteria, Institute of Entomology, Czechoslovak Academy of Sciences, Flemingovo nám. 2, Praha 6, Czechoslovakia.
CCM = Czechoslovak Collection of Microorganisms, J.E. Purkyne University, 662 43 BRNO, Tr. Obráncu míru 10, Czechoslovakia.
F = Food Hygiene Laboratory, Central Public Health Laboratory, Colindale Avenue, London NW9 5HT, England.
IAM = Institute of Applied Microbiology, University of Tokyo, Bunkyo-ku, Tokyo, Japan.
NCIB = National Collection of Industrial Bacteria, PO Box 31, Torry Research Station, 135 Abbey Road, Aberdeen AB9 8DG, Scotland.
NCTC = National Collection of Type Cultures, Central Public Health Laboratory, Colindale Avenue, London NW9 5HT, England.
NRRL = Northern Regional Research Laboratories, Peoria, Illinois, USA.

B. anthracis A19432/80 (**Figures 21**, **22A**, **23A**, **24** and **43**) was from the South African Institute for Medical Research, Johannesburg, SA.

B. anthracis (no number) in **Figure 42** was an isolate submitted to the Public Health Laboratory, Guildford.

B. cereus var. *terminalis* ('T-strain') – **Figure 32** – was supplied by Dr F.F. Busta, Department of Food Science and Industries, University of Minnesota, St Paul, Minnesota 55101, USA.

B. thuringiensis strains were kindly supplied by Professor J.R. Norris, Meat Research Institute, Langford, Bristol.

Media and reagents

Ammonium oxalate–crystal violet

Solution A:

Crystal violet	10.0 g
Ethanol (95%)	100 ml

Mix and dissolve.

Solution B:

Ammonium oxalate	1% aqueous solution

For use, mix 20 ml of solution A and 80 ml of solution B.

Ammonium salt sugars

$(NH_4)_2HPO_4$	1.0 g
KCl	0.2 g
$MgSO_4.7H_2O$	0.2 g
Yeast extract	0.2 g
Agar (Davis, New Zealand)	13.0 g
Distilled water	1000 ml

Add the solids to the distilled water and dissolve by steaming. Add 3.2 ml of 1% aqueous bromothymol blue, adjust the pH to 7.1 and distribute into 100 ml volumes. Sterilize at 115°C for 20 min.

The appropriate carbohydrate is added as a filter sterilized concentrate to give a final concentration of 1%. Mix and distribute into tubes and slope.

Anaerobic agar

(Available commercially from BBL – No. 10929)

Trypticase	20.0 g
NaCl	5.0 g
Agar	15.0 g
Sodium thioglycollate	2.0 g
Sodium formaldehyde sulphoxylate	1.0 g
Distilled water	1000 ml

Dissolve the ingredients in the water with heating, adjust the pH to 7.2 and sterilize at 115°C for 20 min.

Bicarbonate agar (Burdon, 1956)

(For *in vitro* production of encapsulated *B. anthracis*)

Dehydrated nutrient broth (Difco)	0.8%
Yeast extract (Difco)	0.3%
Glucose	0.5%
Agar	2.5%

The medium, made up with gentle heating in distilled water is sterilized by autoclaving at 115°C for 10 min. A 7% solution of $NaHCO_3$, sterilized by filtration is added to the sterile molten agar medium to give a final concentration of 0.7% bicarbonate.

Sterile bovine serum albumin fraction V to a final concentration of 0.7% w/v or human, sheep, calf or horse serum (in that order of preference) to a final concentration of 20% v/v should then be added.

Buffalo black (Smirnoff, 1962)

Buffalo black powder (Sigma stains N/9002, practical grade)	1.5 g
98% methanol	50 ml
Glacial acetic acid	10 ml
Distilled water	40 ml

Dissolve the dye powder in the mixture of liquids and store at least 3 days before use.

Carbol fuchsin (1%)

Dissolve 10 g of basic fuchsin in 100 ml of absolute ethanol and add to 1000 ml of 5% aqueous phenol.

Dilute further using distilled water as required for 0.1% carbol-fuchsin.

Casein agar

Oxoid Skim Milk Powder (L31) made up at double strength.

Davis, New Zealand agar (molten at 50°C) at double strength. Equal quantities of the double strength solutions are mixed at 50°C and poured into Petri dishes.

Christensen's urea medium

Peptone (Oxoid L34)	1.0 g
NaCl	5.0 g
KH_2PO_4	2.0 g
Agar	20.0 g
Distilled water	1000 ml

Dissolve the ingredients by gentle heating and adjust pH to 6.8. Filter and sterilize at 115°C for 20 min.

Glucose	1.0 g
0.2% solution phenol red	6 ml

Add the glucose and indicator to the cooled molten base and steam for 1 hour. Cool to 50–55°C and add 100 ml of a filter sterilized 20% solution of urea to the base aseptically. Distribute into test tubes and slope.

Columbia blood agar

Available commercially from Oxoid – No. CM331. Sterile defibrinated horse blood is added aseptically to the prepared base (cooled to 50°C) to give a final concentration of 5% blood agar.

Craigie agar

Nutrient broth No. 2 (Oxoid CM 67)	25.0 g
Agar (Oxoid No. 1, L11)	2.5 g
Distilled water	1000 ml

Mix the ingredients and bring to the boil with constant stirring. Distribute into wide test tubes or McCartney bottles with Craigie tubes. Sterilize at 115°C for 20 min.

Defined complex amino acid medium
(Proom and Knight, 1955)

A. Ammonia basal salt medium

KH_2PO_4	1.500 g
$(NH_4)_2HPO_4$	7.000 g
$MgSO_4.7H_2O$	0.500 g
$CaCl_2.2H_2O$	0.300 g
$MnSO_4.4H_2O$	0.040 g
$FeSO_4.7H_2O$	0.025 g
Ammonium molybdate	0.002 g

Dissolve in 1000 ml of distilled water and adjust the pH to 7.6. Boil, pass through filter paper and sterilize at 115°C for 20 min.

B. Complex amino acid medium

The ammonia basal salt medium is supplemented with the following amino acids in the concentrations shown (mg/ml): DL-alanine, 0.38; DL-aspartic acid, 0.89; L-arginine HCl, 0.3; L-cystine, 0.02; L-glutamic acid, 1.4; glycine, 0.17; L-histidine HCl, 0.24; DL-isoleucine, 0.76; L-leucine, 0.57; L-lysine HCl, 0.24; DL-methionine, 0.06; L-proline, 0.11; DL-serine, 0.12; DL-threonine, 0.10; L-tyrosine, 0.06; DL-valine, 0.15.

The amino acid mixture is prepared separately by solution in distilled water at 10 times the required concentration, and sterilized by Seitz filtration. It is then added aseptically to the ammonia basal medium to give the required concentration.

Diagnostic Sensitivity Test (DST) Agar

Available commercially from Oxoid – No. CM261.

Ehrlich's reagent

p-dimethylaminobenzaldehyde	1.0 g
Absolute ethanol	95 ml
Concentrated HCl	20 ml

Dissolve the aldehyde in the ethanol and add the acid. Protect from light.

Giemsa stain

Stock solution:

Giemsa stain powder	1.0 g
Glycerol	60 ml
Absolute methanol	60 ml

Heat the glycerol to 55–60°C in a water bath. Add the stain and mix thoroughly. Incubate the mixture at the same temperature for 2 hours shaking periodically to help the stain dissolve. Allow to cool and add the methanol. It is best used after about 2 weeks' 'maturation' on the shelf.

For use: Dilute the stain 1:10 and insert the slide for 30 min to 1 hour. Wash with distilled water, dry and examine.

Glucose phosphate broth

Peptone (Oxoid L37)	5.0g
K_2HPO_4	5.0g
Distilled water	1000ml

Steam to dissolve the solids and adjust pH to 7.5. Add 5.0g of glucose, dissolve, mix and distribute the solution into test tubes. Sterilize at 115°C for 10 min.

Hydrogen peroxide (catalase activity)

H_2O_2 3% aq. soln ('10 volume').

Protect from light and store in a cool place. Keep in a bottle with a glass stopper or plastic screw cap.

'J' agar

Tryptone	5.0g
Yeast extract	15.0g
K_2HPO_4	3.0g
Agar	20.0g
Tap-water	1000ml

Mix the ingredients and adjust the pH to 7.3–7.5. Sterilize at 121°C for 20 min.

Glucose 2.0g as a filter sterilized concentrate

Add the sterile glucose solution aseptically after autoclaving and cooling to 50°C. Pour into Petri dishes.

'J' medium

A semi-solid variation of 'J' agar having only 0.1% agar. No glucose is added. The medium is poured into test tubes or screw-capped bottles.

'J' medium – diphasic

Double-strength 'J' agar base without the agar is aseptically superimposed on 4% agar prepoured and allowed to set on the bottom of a conical flask. The volumes of agar and liquid base should be approximately equal.

Kendall's BC Medium

(Miss M. Kendall, Food Hygiene Laboratory)

Base

Ammonium hydrogen phosphate	1.0 g
KCl	0.2 g
$MgSO_4.4H_2O$	0.2 g
Yeastrel	0.2 g
Agar	20.0 g
Distilled water	1000 ml

Mix ingredients and steam to dissolve. Filter, adjust medium to pH 7 and add indicator.

1% alc. bromocresol purple	4 ml

Distribute the base in 90 ml quantities and sterilize for 10 min at 115°C.

Egg yolk: Separate eggs aseptically, break the yolks and mix with an equal volume of physiological saline. Filter through gauze, distribute in sterile bottles and heat for 30 min at 60°C.

Store yolk emulsion at 4°C until required.

For use: To 90 ml of base, add 1.0 g of mannitol. Melt in a steamer and allow to cool at 50°C. Add 10 ml of the egg yolk emulsion, mix well and pour into Petri dishes.

Kovács' reagent

p-dimethylaminobenzaldehyde	5.0 g
Amyl alcohol	75 ml
Concentrated HCl	25 ml

Dissolve the aldehyde in the alcohol by gentle warming in a water bath (50–55°C). Cool and add the acid. Protect from light (wrap the bottle with aluminium foil, for instance) and store at 4°C.

Lactose broth

This is nutrient broth with added 1% lactose.

Lugol's iodine

Iodine	5.0 g
Potassium iodide	10.0 g
Distilled water	100 ml

Dissolve the KI and iodine in 10 ml of the water and add the remaining 90 ml of water. For use, dilute 1:5 with distilled water.

0.001% Lysozyme

Prepare a solution of 0.1 g of lysozyme in 65 ml sterile 0.01 N HCl. Boil for 20 min, cool to ambient temperature and make up to 100 ml with further 0.01 N HCl. Mix 1 ml of this lysozyme solution with 99 ml of sterile nutrient broth and dispense 2.5 ml volumes into sterile tubes.

Malonate broth

$(NH_4)_2SO_4$	2.0 g
K_2HPO_4	0.6 g
KH_2PO_4	0.4 g
NaCl	2.0 g
Sodium malonate	3.0 g
DL-phenylalanine	2.0 g
Yeast extract	1.0 g
Distilled water	1000 ml
Bromothymol blue (0.2% solution)	12.5 ml

Dissolve the solids in the water by heating. Filter, add the indicator and distribute into test tubes. Sterilize at 115°C for 20 min.

Methylene blue medium (Burdon, 1956)

(For observation of methylene blue reduction)

Tryptose	20.0 g
NaCl	5.0 g
Agar	8.0 g
Distilled water	1000 ml

After autoclaving, the medium is cooled to 55°C and a sufficient amount of 1% aqueous methylene blue added to give a faint but distinct blue colour. The completed medium is adjusted to a pH of 7.2–7.4 and distributed aseptically into test tubes to give an agar column several inches deep.

Inoculations are made with a single stab using a straight wire.

Mueller–Hinton agar

(Available commercially from Difco – No. B252.)

α-Naphthol solution (V-P test)

Make up a 5% solution of α-naphthol in absolute ethanol. Store in the refrigerator in a dark bottle.

Nitrate broth

KNO_3	1.0g
Nutrient broth No. 2 (Oxoid CM 67)	1000ml

Dissolve the KNO_3 in the broth, distribute into tubes containing inverted Durham's tubes, and sterilize at 115°C for 20 min.

Nitrate reagents A & B

A: 0.8% sulphanilic acid in 5N acetic acid
B: 0.5% alpha-naphthylamine in 5N acetic acid

Dissolve both (separately) by gentle heating. Store at room temperature but protect reagent A from light (wrap the container in aluminium foil, for example). Note: the chemicals are potentially carcinogenic and should be handled with caution.

Nutrient broth and nutrient agar

Beef extract	10.0g	or Nutrient broth No. 2 (Oxoid CM 67)
Peptone	10.0g	
NaCl	5.0g	
Distilled water	1000ml	

Dissolve ingredients by heating. Adjust pH to 8–8.4 and boil for 10 min. Filter, adjust pH 7.2–7.4, distribute into test tubes in required volumes and sterilize 115°C for 20 min. Nutrient agar is obtained by adding 12g of agar (Davis, New Zealand) to recipe.

Nutrient gelatin

Beef extract	3.0g
Peptone	5.0g
Gelatin (Bacto nutrient gelatin [B11] Difco)	120.0g
Distilled water	1000ml

Add the gelatin to the water and allow to stand for 15–30 min. Heat to dissolve the gelatin; add and dissolve the other ingredients. Adjust the pH to 7.0 and distribute to test tubes. Sterilize at 115°C for 20 min.

Peptone water

Peptone (Lab M peptone No. I)	10.0 g
NaCl	5.0 g
Distilled water	1000 ml

Boil the water and add the ingredients. Stir until dissolved. Allow to cool and adjust the pH to 7.6. Distribute in the required volumes and containers and sterilize at 115°C for 20 min.

PLET medium (Knisely, 1966)

Heart infusion agar (Difco) is made up in the proportions recommended by Difco, the pH adjusted to 7.35, autoclaved and cooled to 50°C. To this molten agar is added 30 units/ml polymyxin, 40 μg/ml lysozyme, 200 μg/ml EDTA and 40 μg/ml thallous acetate.

Polychrome methylene blue

Prepare a saturated solution of methylene blue in 95% ethanol by mixing ca. 0.5 g of the dye in 50 ml of the alcohol. Add 30 ml of this to 100 ml of a 0.01% KOH solution in distilled water. Add K_2CO_3 to a final concentration of 1%.

Allow to stand in half-filled bottles with light cotton-wool plugs. The bottles should be shaken periodically for fuller aeration. Oxidation ('ripening') of the methylene blue occurs slowly, taking several months. It is marked by the development of a violet compound giving the stain its polychrome properties.

Stain smears for about 5 min and wash off with water.

Potato starch agar

Potato starch	9.5 g
Distilled water	1000 ml
Nutrient broth No. 2 (Oxoid CM 67)	25.0 g
Davis, New Zealand agar	12.0 g

Add the starch to 100 ml of the distilled water. Dissolve the 25 g of nutrient broth powder in the remaining water. Add the agar to the nutrient broth and boil to dissolve. Add the starch solution to the molten nutrient agar and mix well. Sterilize at 115°C for 10 min and pour plates.

Propionate utilization (Gordon *et al*, 1973)

This is a modification of Koser's citrate agar.

NaCl	1.0 g
$MgSO_4.7H_2O$	0.2 g
$(NH_4)_2HPO_4$	0.5 g
Agar	15.0 g
Sodium propionate	2.0 g
Distilled water	1000 ml
0.04% (w/v) solution phenol red	20 ml

Mix the ingredients, adjust pH to 6.8 before sterilizing as for Simmons' citrate. Slope in test tubes.

Salt trypticase broth (7%)

Dissolve 10 g of trypticase (BBL 11921) in 1000 ml of distilled water with gentle heating and add 70 g of NaCl. Stir until dissolved, distribute into tubes or screw-capped bottles and sterilize at 115°C for 10 min.

Simmons' citrate

NaCl	5.0 g	or Oxoid CM 155 powder
$MgSO_4.7H_2O$	0.2 g	
$NH_4H_2PO_4$	1.0 g	
K_2HPO_4	1.0 g	
Distilled water	1000 ml	

Dissolve the salts in the water and add 2 g of citric acid. Adjust the pH to 6.8 with 1 N NaOH and filter through a sintered glass funnel. Add 40 ml of a 0.2% solution of bromothymol blue and 20 g of agar (Davis, New Zealand). Sterilize at 115°C for 20 min.

Sporulation agar

Nutrient broth No. 2 (Oxoid CM 67)	6.0 g
$MnSO_4.4H_2O$	0.03 g
KH_2PO_4	0.25 g
Agar No. 3	12.0 g
Distilled water	1000 ml

Mix ingredients and dissolve by heating. Sterilize at 115°C for 20 min. Distribute either in plates or tubes (slopes).

Soil-extract agar

Garden soil, air-dried	1000 g
Distilled water	2400 ml

Sift the air-dried soil through a fine sieve and add to the water. Mix well and autoclave for 1 hour at 121°C. Stir and filter through filter paper. If the filtrate is turbid, it may be clarified by addition of talc followed by refiltering.

To make the soil extract agar:

Peptone	5.0 g
Beef extract	3.0 g
Agar	20.0 g
Soil extract from above	1000 ml

Add the ingredients to the extract and heat to dissolve. Adjust the pH to 7.0 and sterilize at 115°C for 20 min.

Tyrosine agar

Suspend 0.5 g of L-tyrosine in 10 ml of distilled water. Autoclave and mix with 100 ml of sterile nutrient agar (molten). Cool to 50°C and pour into Petri dishes. (Care must be taken to ensure even distribution of tyrosine crystals in the medium by thorough mixing before pouring.)

References

Ainsworth, G.C., Brown, A.M. and Brownlee, G. (1947). 'Aerosporin', an antibiotic produced by *Bacillus aerosporus* Greer. *Nature* **160**, 263.

Allen. B.T. and Wilkinson, H.A. (1969). A case of meningitis and generalized Shwartzman reaction caused by *Bacillus sphaericus*. *Johns Hopkins Medical Journal* **125**, 8.

Approved Lists of Bacterial Names (1980). *International Journal of Systematic Bacteriology* **30**, 256.

Ascoli, A. (1911). Die Präzipitindiagnose bei Milzbrand. *Zentralblatt für Bakteriologie, Parasitenkunde und Infektionskrankheiten* **58**, 63.

ATCC Catalogue of Strains. I. (1982). 15th ed. American Type Culture Collection, 12301, Parklawn Drive, Rockville, Maryland 20852, USA.

Bailey, L. (1981). *Honey Bee Pathology*. Academic Press, London.

Bailie, W.E. and Stowe, E.C. (1977). A simplified test for identification of *Bacillus anthracis*. *Abstracts of the Annual Meeting of the American Society for Microbiology*, C80, p.48.

Baker, F.J. and Breach, M.R. (1980). *Medical Microbiological Techniques*, pp.156–159. Butterworth, London.
de Barjac, H. (1981). Insect pathogens in the genus *Bacillus*. In: *The Aerobic Endospore – forming Bacteria: Classification and Identification*, ed. by R.C.W. Berkeley and M. Goodfellow, pp.241–250. Academic Press, London.
de Barjac, H. and Bonnefoi, A. (1962). Essai de classification biochimique et serologique de 24 souches de *Bacillus* de type *B. thuringiensis*. *Entomophaga* **8**, 223.
Bechtel, D.B. and Bulla, L.A. (1976). Electron microscope study of sporulation and parasporal crystal formation in *Bacillus thuringiensis*. *Journal of Bacteriology* **127**, 1472.
Belton, F.C. and Strange, R.E. (1954). Studies on a protective antigen produced *in vitro* from *Bacillus anthracis*: medium and methods of production. *British Journal of Experimental Pathology* **35**, 144.
Berkeley, R.C.W. and Goodfellow, M. (1981). *The Aerobic Endospore – forming Bacteria: Classification and Identification.* Special Publication of the Society for General Microbiology. Academic Press, London.
van Bijsterveld, O.P. and Richards, R.D. (1965). *Bacillus* infections of the cornea. *Archives of Ophthalmology* **74**, 91.
Block, C.S., Levy, M.L. and Fritz, V.U. (1978). *Bacillus cereus* endocarditis. A case report. *South African Medical Journal* **53**, 556.
Bonventre, P.F. and Johnson, C.E. (1970). *Bacillus cereus* toxin. In: *Microbial Toxins, 3, Bacterial Protein Toxins,* ed. by T.C. Montie, S. Kadis and S.J. Ajl, pp.415–435. Academic Press, London.
Borick, P.M. and Fogarty, M.G. (1967). Effects of continuous and interrupted radiation on microorganisms. *Applied Microbiology*. **15**, 785.
Boyette, D.P. and Rights, F.L. (1952). Heretofore undescribed aerobic spore bearing bacillus in child with meningitis. *Journal of the American Medical Association* **148**, 1223.
Bradley, D.E. and Franklin, J.G. (1958). Electron microscope survey of the surface configuration of spores of the genus *Bacillus*. *Journal of Bacteriology* **76**, 618.
Brown, E.R. and Cherry, W.B. (1955). Specific identification of *Bacillus anthracis* by means of a variant bacteriophage. *Journal of Infectious Diseases* **96**, 34.
Brown, E.R., Moody, M.D., Treece, E.L. and Smith, C.W. (1958). Differential diagnosis of *Bacillus cereus*, *Bacillus anthracis*, and *Bacillus cereus* var *mycoides*. *Journal of Bacteriology* **75**, 499.
Buck, C.A., Anacker, R.L., Newman, F.S. and Eisenstark, A. (1963). Phage isolated from lysogenic *Bacillus anthracis*. *Journal of Bacteriology* **85**, 1423.
Burdon, K.L. (1956). Useful criteria for the identification of *Bacillus anthracis* and related species. *Journal of Bacteriology* **71**, 25.

Burges, H.D. (personal communication). Dr H.D. Burges, Insect Pathology Group, Glasshouse Crops Research Institute, Worthing Road, Rustington, Littlehampton, West Sussex BN16 2PU.
Burges, H.D. (1981). Safety, safety testing and quality control of microbial pesticides. In: *Microbial Control of Pests and Plant Diseases, 1970–1980,* ed. by H.D. Burges, pp.737–767. Academic Press, New York.
Burges, H.D. (1982). Control of insects by bacteria. *Parasitology* **84**, 79.
Candeli, A., de Bartolomeo, A., Mastrandea, V. and Trotta, F. (1979). Contribution to the characterization of *Bacillus megaterium. International Journal of Systematic Bacteriology* **29**, 25.
Capel, B.J. (personal communication). Mr B.J. Capel, Vaccine Research and Production Laboratory, PHLS–CAMR, Porton Down, Salisbury, Wiltshire SP4 0JG.
Carman, J.A. (personal communication). Mr J.A. Carman, Vaccine Research and Production Laboratory, PHLS–CAMR, Porton Down, Salisbury, Wiltshire SP4 0JG.
Charlton, R. (1980). Anthrax and the string of pearls. *Zimbabwe Medical Technology Journal* **9**, 71.
Charlton, R. (personal communication). Mr R. Charlton, Bulawayo Group Laboratory, PO Box 2096, Bulawayo, Zimbabwe.
Cherry, W.B. and Freeman, E.M. (1959). Staining bacterial smears with fluorescent antibody. *Zentralblatt für Bakteriologie, Parasitenkunde, Infektionskrankheiten und Hygiene* **175**, 582.
Colmer, A.R. (1947). The use of the enzyme lecithinase in grouping some members of the genus *Bacillus*. *Journal of Bacteriology* **54**, 11.
Cowan, S.T. (1974). *Cowan and Steel's Manual for the Identification of Medical Bacteria.* Cambridge University Press, Cambridge.
Cruickshank, R., Duguid, J.P., Marmion, B.P. and Swain, R.H.A. (1975). *Medical Microbiology*, 12th ed., Vol. 2, pp.33 and 449–453. Churchill Livingstone, Edinburgh.
Dancer, B.N. (personal communication). Dr B.N. Dancer, Microbiology Unit, Department of Biochemistry, University of Oxford, South Parks Road, Oxford OX1 3QU.
Darmady, E.M., Hughes, K.E.A., Burt, M.M., Freeman, B.M. and Powell, D.B. (1961). Radiation sterilization. *Journal of Clinical Pathology* **14**, 55.
Davies, J.C.A. (personal communication). Dr J.C.A Davies, 2 Aboyne Drive, Highlands, Harare, Zimbabwe.
Dienes, L. (1946). Reproductive processes in *Proteus* cultures. *Proceedings of the Society for Experimental Biology and Medicine* **63**, 265.
Elter, V.B. (1966). Veitrag zum problem der lebensmittelvergiftungen durch aerobe sporenbildner. *Zeitschrifft für die gesampte Hygiene und ihre Grenzgebiete* **12**, 65.

English, C.F., Bell, E.J. and Berger, A.J. (1967). Isolation of thermophiles from broadleaf tobacco and effect of pure culture inoculation on cigar aroma and mildness. *Applied Microbiology* **15**, 117.

Falcon, L.A. (1971). Use of bacteria for microbial control of insects. In: *Microbial Control of Insects and Mites*, ed. by H.D. Burges and N.W. Hussey, pp.67–83. Academic Press, London.

Farrar, W.E. (1963). Serious infections due to 'non-pathogenic' organisms of the genus *Bacillus*. *American Journal of Medicine* **34**, 134.

Feeley, J.C. and Patton, C.M. (1980). *Bacillus*. In: *Manual of Clinical Microbiology*, ed. by E.H. Lennette, A. Balows, W.J. Hausler and J.P. Truant, 3rd ed. American Society for Microbiology, Washington, D.C.

Fitzpatrick, D.J., Turnbull, P.C.B., Keane, C.T. and English, L.F. (1979). Two gas-gangrene-like infections due to *Bacillus cereus*. *British Journal of Surgery* **66**, 577.

Garrod, L.P., Lambert, H.P. and O'Grady, F. (1981). *Antibiotic and Chemotherapy*, 5th ed. Churchill Livingstone, Edinburgh.

Gibson, T. and Gordon, R.E. (1974). Genus *Bacillus*. In: *Bergey's Manual of Determinative Bacteriology*, ed. by R.E. Buchanan and N.E. Gibbons, 8th ed., pp.529–550. Williams and Wilkins, Baltimore.

Gilbert, R.J. (1979). *Bacillus cereus* gastroenteritis. In: *Foodborne Infections and Intoxications*, ed. by H. Riemann and F.L. Bryan, 2nd ed., pp. 495–518. Academic Press, New York.

Gilbert, R.J. and Parry, J.M. (1977). Serotypes of *Bacillus cereus* from outbreaks of food poisoning and from routine foods. *Journal of Hygiene, Cambridge* **78**, 69.

Gilbert, R.J. and Taylor, A.J. (1976). *Bacillus cereus* food poisoning. In: *Microbiology in Agriculture, Fisheries and Food*, ed. by F.A. Skinner and J.C. Carr, pp.197–213. Society for Applied Bacteriology, Symposium Series No. 4. Academic Press, London.

Gilbert, R.J., Turnbull, P.C.B., Parry, J.M. and Kramer, J.M. (1981). *Bacillus cereus* and other *Bacillus* species: their part in food poisoning and other clinical infections. In: *The Aerobic Endospore–forming Bacteria: Classification and Identification*, ed. by R.C.W. Berkeley and M. Goodfellow, pp.297–314. Academic Press, London.

Gordon, R.E. (1975). Washington's hometown project on the genus *Bacillus*. *ASM News* **41**, 715.

Gordon, R.E. (1981). One hundred and seven years of the genus *Bacillus*. In: *The Aerobic Endospore–forming Bacteria: Classification and Identification*, ed. by R.C.W. Berkeley and M. Goodfellow, pp.1–15. Academic Press, London.

Gordon, R.E., Haynes, W.C. and Pang, C.H-N. (1973). The genus *Bacillus*. United States Department of Agriculture, Agricultural Research Service, Agriculture Handbook No. 427. Washington, D.C. (Sold by the Superintendent of Documents, U.S. Government Printing

Office, Washington, D.C. 2040.)

Hunger, W. and Claus, D. (1981). Taxonomic studies on *Bacillus megaterium* and on agarolytic *Bacillus* strains. In: *The Aerobic Endospore – forming Bacteria: Classification and Identification*, ed. by R.C.W. Berkeley and M. Goodfellow, pp.217–239. Academic Press, London.

Ihde, D.C. and Armstrong, D. (1973). Clinical spectrum of infection due to *Bacillus* species. *American Journal of Medicine* **55**, 839.

Ingram, M. (1969). Sporeformers as food spoilage organisms. In: *The Bacterial Spore*, ed. by G.W. Gould and A. Hurst, pp.549–610. Academic Press, London.

Isaacson, P., Jacobs, P.H., Mackenzie, A.M.R. and Mathews, A.W. (1976). Pseudotumour of the lung caused by infection with *Bacillus sphaericus*. *Journal of Clinical Pathology* **29**, 806.

Jensen, J. and Kleemeyer, H. (1953). Die bakterielle differentialdiagnose des anthrax mittels eines neuen spezifischen testes ('perlschnurtest'). *Zentralblatt für Bacteriologie, Parasitenkunde, Infektionskrankheiten und Hygiene* **159**, 494.

Johnston, K.G. (1981). Bovine mastitis caused by *Bacillus cereus*. *Veterinary Record* **108**, 404.

Johnston, K.G. (personal communication). Dr K.G. Johnston, Senior Clinical Pathologist, University of Sydney, Rural Veterinary Centre, Werombi Road, Camden, N.S.W. 3570, Australia.

Jones, T.O. and Turnbull, P.C.B. (1981). Bovine mastitis caused by *Bacillus cereus*. *Veterinary Record* **108**, 271.

Jordan, E.O. (1890). A report on certain species in bacteria observed in sewage. Cited by Gordon *et al*, 1973.

Kaneko, T., Nozaki, R. and Aizawa, K. (1978). Deoxyribonucleic acid relatedness between *Bacillus anthracis*, *Bacillus cereus* and *Bacillus thuringiensis*. *Microbiology and Immunology (Tokyo)* **22**, 639.

Kavanagh, F. (1972). *Analytical Microbiology*. Vol. 2, pp.308–312. Academic Press, New York.

Klein, M.G. (1981). Advances in the use of *Bacillus popilliae* for pest control. In: *Microbial Control of Pests and Plant Diseases, 1970–1980*, ed. by H.D. Burges, pp.183–192. Academic Press, New York.

Knight, B.C.J.G. and Proom, H. (1950). A comparative survey of the nutrition and physiology of mesophilic species in the genus *Bacillus*. *Journal of General Microbiology* **4**, 508.

Knisely, R.F. (1966). Selective medium for *Bacillus anthracis*. *Journal of Bacteriology* **92**, 784.

Knox, R. and Collard, P. (1952). The effect of temperature on the sensitivity of *Bacillus cereus* to penicillin. *Journal of General Microbiology* **6**, 369.

Koch, R., Gaffky, G.T.A. and Loeffler, F.A.J. (1884). Cited in *Topley*

and Wilson's Principles of Bacteriology, Virology and Immunity, revised by G.S. Wilson and A.A.Miles, 6th ed., Vol. 2, p.2216. Edward Arnold, London.
Kramer, J.M., Turnbull, P.C.B., Munshi, G. and Gilbert, R.J. (1982). Identification and characterization of *Bacillus cereus* and other *Bacillus* species associated with foods and food poisoning. In: *Methods for the Isolation and Identification of Food Poisoning Organisms*, ed. by J.E.L. Corry, D. Roberts and F.A. Skinner, pp.261–286. Society for Applied Bacteriology, Technical Series, No. 17. Academic Press, London.
Lemille, F., Barjac, H. de and Bonnefoi, A. (1969). Étude sérologique de *Bacillus cereus*. Mise en évidence de divers sérotypes basés sur les antigènes flagellaires. *Annales Institut Pasteur (Paris)* **117**, 31.
Lillie, R.D. (1928). The Gram stain. 1. A quick method for staining Gram-positive organisms in the tissues. *Archives of Pathology* **5**, 828.
Logan, N.A. and Berkeley, R.C.W. (1981). Classification and identification of members of the genus *Bacillus* using API tests. In: *The Aerobic Endospore – forming Bacteria: Classification and Identification*, ed. by R.C.W. Berkeley and M. Goodfellow, pp.105–140. Academic Press, London.
McCloy, E.W. (1951a). Studies on a lysogenic bacillus strain. I. A bacteriophage specific for *Bacillus anthracis*. *Journal of Hygiene* **49**, 114.
McCloy, E. (1951b). Unusual behaviour of a lysogenic *Bacillus* strain. *Journal of General Microbiology* **5**, xiv.
McCloy, E.W. (1958). Lysogenicity and immunity to *Bacillus* phage W. *Journal of General Microbiology* **18**, 198.
McClung, L.S. and Toabe, R. (1947). The egg yolk reaction for the presumptive diagnosis of *Clostridium sporogenes* and certain species of the gangrene and botulinum groups. *Journal of Bacteriology* **53**, 139.
Melles, Z., Nikodemusz, I. and Abel, A. (1969). Die pathogene Wirkung aerober sporenbildender Bakterien. *Zentralblatt für Bakteriologie, Parasitenkunde, Infektionskrankheiten und Hygiene* **212**, 174.
Meynell, E. and Meynell, G.G. (1964). The roles of serum and carbon dioxide in capsule formation by *Bacillus anthracis*. *Journal of General Microbiology* **34**, 153.
M'Fadyean, J. (1903a). A peculiar staining reaction of the blood of animals dead of anthrax. *Journal of Comparative Pathology* **16**, 35.
M'Fadyean, J. (1903b). A further note with regard to the staining reaction of anthrax blood with methylene blue. *Journal of Comparative Pathology* **16**, 360.
M'Fadyean, J. (1904). The colour reaction of anthrax blood with methylene blue: a question of priority of publication. *Journal of Comparative Pathology* **17**, 58.
Milner, R.J. (1981). Identification of the *Bacillus popilliae* group of insect pathogens. In: *Microbial Control of Pests and Plant Diseases,*

1970–1980, ed. by H.D. Burges, pp.45–59. Academic Press, New York.

Nordberg, B.K. (1953). Continued investigations of some important characteristics in anthrax-like microorganisms as viewed from a point of view of differential diagnosis. *Nordisk Veterinaer Medicin* **5**, 915.

Norris, J.R., Berkeley, R.C.W., Logan, N.A. and O'Donnell, A.G. (1981). The genera *Bacillus* and *Sporolactobacillus*. In: *The Prokaryotes: a Handbook on Habitats, Isolation, and Identification of Bacteria*, ed. by M.P. Starr, H. Stolp, H.G. Trüper, A. Balows and H.G. Schlegel, Vol. 2, pp.1711–1742. Springer–Verlag, Berlin.

O'Day, D.M., Smith, R.S., Gregg, C.R., Turnbull, P.C.B., Head, W.S., Ives, J.A. and Ho, P.C. (1981). The problem of *Bacillus* species infection with special emphasis on the virulence of *Bacillus cereus*. *Ophthalmology* **88**, 833.

Oppenheim, B.A. and Koornhof, H.J. (1980). A selective medium for anthrax. Proceedings of the Congress of the South African Society of Pathologists, 7–9 July, 1980, South African Institute for Medical Research, Johannesburg.

van Os, J.L., Lameris, S.A., Doodeward, J. and Oostendorp, J.G. (1975). Diffusion test for the determination of antibiotic residues in milk. *Netherlands Milk and Dairy Journal* **29**, 16.

Ouderkirk, L.A. (1979). *Bacillus stearothermophilus* disk assay for detection of residual penicillins in milk: collaborative study. *Journal of the Association of Official Analytical Chemists* **62**, 985.

Pasteur, L. (1881a and b). Cited in *Topley and Wilson's Principles of Bacteriology, Virology and Immunity*, revised by G.S. Wilson and A.A. Miles, 6th ed., Vol. 1, p.1091. Edward Arnold, London.

Prasad, S.S.S.V. and Shethna, Y.I. (1974). Purification, crystallization and partial characterization of the antitumour and insecticidal protein subunit from the delta–endotoxin of *Bacillus thuringiensis* var. *thuringiensis*. *Biochimica et Biophysica Acta* **363**, 558.

Prasad, S.S.S.V. and Shethna, Y.I. (1975). Enhancement of immune response by the proteinaceous crystal of *Bacillus thuringiensis* var. *thuringiensis*. *Biochemical and Biophysical Research Communications* **62**, 517.

Priest, F.G. (1981). DNA homology in the genus *Bacillus*. In: *The Aerobic Endospore – forming Bacteria: Classification and Identification*, ed. by R.C.W. Berkeley and M. Goodfellow, pp.33–57. Academic Press, London.

Priest, F.G., Goodfellow, M. and Todd, C. (1981). The genus *Bacillus*: a numerical analysis. In: *The Aerobic Endospore – forming Bacteria: Classification and Identification*, ed. by R.C.W. Berkeley and M. Goodfellow, pp.91–103. Academic Press, London.

Proom, H. and Knight, B.C.J.G. (1955). The minimal nutritional

requirements of some species in the genus *Bacillus*. *Journal of General Microbiology* **13**, 474.
Schaeffer, A.B. and Fulton, M. (1933). A simplified method of staining endospores. *Science* **77**, 194.
Seenappa, M. and Kempton, A.G. (1981). A note on the occurrence of *Bacillus cereus* and other species of *Bacillus* in Indian spices of export quality. *Journal of Applied Bacteriology* **50**, 225.
Singer, S. (1981). Potential of *Bacillus sphaericus* and related spore-forming bacteria for pest control. In: *Microbial Control of Pests and Plant Diseases, 1970–1980*, ed. by H.D. Burges, pp.183–298. Academic Press, New York.
Smirnoff, W.A. (1962). A staining method for spores, crystals and vegetative cells of *Bacillus thuringiensis* Berliner. *Journal of Insect Pathology* **4**, 384.
Smith, N.R., Gordon, R.E. and Clark, F.E. (1946). Aerobic mesophilic spore-forming bacteria. U.S.D.A. Miscellaneous Publication 559.
Smith, N.R., Gordon, R.E. and Clark, F.E. (1952). Aerobic spore-forming bacteria. U.S. Department of Agriculture, Agriculture Monograph, No. 16.
Sterne, M. (1937). Variation in *Bacillus anthracis*. *Onderstepoort Journal of Veterinary Science and Animal Industry* **8**, 271.
Sterne, M. (1939). The immunization of laboratory animals against anthrax. *Onderstepoort Journal of Veterinary Science and Animal Industry* **13**, 313.
Stratford, B.C. (1977). *An Atlas of Medical Microbiology: Common Human Pathogens*. pp.73–74. Blackwell Scientific Publications, Oxford.
Tabbara, K.F. and Tarabay, N. (1979). *Bacillus licheniformis* corneal ulcer. *American Journal of Ophthalmology* **87**, 717.
Taylor, A.J. and Gilbert, R.J. (1975). *Bacillus cereus* food poisoning: a provisional serotyping scheme. *Journal of Medical Microbiology* **8**, 543.
Thorne, C.B., Gomez, C.G. and Housewright, R.D. (1952). Synthesis of glutamic acid and glutamyl polypeptide by *Bacillus anthracis*. *Journal of Bacteriology* **63**, 363.
Tuazon, C.U., Murray, H.W., Levy, C., Solny, M.N., Curtin, J.A. and Sheagren, J.N. (1979). Serious infections from *Bacillus* species. *Journal of the American Medical Association* **241**, 1137.
Turnbull, P.C.B. (1981). *Bacillus cereus* toxins. *Pharmacology and Therapeutics* **13**, 453.
Turnbull, P.C.B., Jørgensen, K., Kramer, J.M., Gilbert, R.J. and Parry, J.M. (1979). Severe clinical conditions associated with *Bacillus cereus* and the apparent involvement of exotoxins. *Journal of Clinical Pathology* **32**, 289.
Turner, M. (1980). Anthrax in humans in Zimbabwe. *Central African*

Journal of Medicine **26**, 160.

Tyrell, D.J., Bulla, L.A., Andrews, R.E., Kramer, K.J., Davidson, L.I. and Nordin, P. (1981). Comparative biochemistry of entomocidal parasporal crystals of selected *Bacillus thuringiensis* strains. *Journal of Bacteriology* **145**, 1052.

Washington, J.A. (1981). *Laboratory Procedures in Clinical Microbiology*. Bacillus. pp.177–178. Springer–Verlag, New York.

Weaver, R.E., Brachman, P.S. and Feeley, J.C. (1970). Anthrax. In: *Diagnostic Procedures for Bacterial, Mycotic and Parasitic Infections*, ed. by H.L. Bodily, E.L. Updike and J.O. Mason, 5th ed., pp.354–363. The American Public Health Association Inc., New York.

Weinstein, L. and Colburn, C.G. (1950). *Bacillus subtilis* meningitis and bacteremia: report of a case and review of the literature on 'subtilis' infections in man. *Archives of Internal Medicine* **86**, 585.

Willemse–Collinet, M.F., Turnbull, P.C.B., Hospers, G.T. and van Oppenray, A.B.W.G. (1981). Computer-assisted method for identification of *Bacillus* species isolated from liquid antacids. *Applied and Environmental Microbiology* **41**, 169.

Williams, R.P. (personal communication). Dr R.P. Williams, Department of Microbiology and Immunology, Baylor College of Medicine, Texas Medical Center, Houston, Texas 77030.

Willis, A.T. and Hobbs, G. (1958). A medium for the identification of clostridia producing opalescence in egg-yolk emulsions. *Journal of Pathological Bacteriology* **75**, 299.

Willis, A.T. and Hobbs, G. (1959). Some new media for the isolation and identification of clostridia. *Journal of Pathological Bacteriology* **77**, 511.

Wolf, J. and Barker, A.N. (1968). The genus *Bacillus*: aids to the identification of the species. In: *Identification Methods for Microbiologists*, part B. pp.93–109, ed. by B.M. Gibbs and D.A. Shapton. Academic Press, London.

Wolf, J. and Sharp, R.J. (1981). Taxonomic and related aspects of thermophiles within the genus *Bacillus*. In: *The Aerobic Endospore – forming Bacteria: Classification and Identification*, ed. by R.C.W. Berkeley and M. Goodfellow, pp.251–296. Academic Press, London.

Wright, C.L. (personal communication). Mr C.L. Wright, MRCVS Veterinary Officer-in-charge, Auchincruivie V.I. Centre, Ayr, Scotland.

Young, E.J., Wallace, R.J., Ericsson, C.D., Harris, R.A. and Clarridge, J. (1980). Panophthalmitis due to *Bacillus cereus*. *Archives of Internal Medicine* **140**, 559.

Index

(References printed in medium type are to page numbers and those in **bold** are to picture and caption numbers)